Sustainable Development and Communication in Global Food Networks

Maria Touri

Sustainable Development and Communication in Global Food Networks

Lessons From India

palgrave
macmillan

Maria Touri
School of Media, Communication and Sociology
University of Leicester
Leicester, UK

ISBN 978-3-030-46118-8 ISBN 978-3-030-46119-5 (eBook)
https://doi.org/10.1007/978-3-030-46119-5

Cover illustration: MirageC / gettyimages
Cover design: eStudioCalamar

This Palgrave Macmillan imprint is published by the registered company Springer Nature Switzerland AG.
The registered company address is: Gewerbestrasse 11, 6330 Cham, Switzerland

Acknowledgements

First and foremost, I would like to thank all the farmers and staff from Organic Wayanad, who facilitated and participated in this research through several rounds of focus groups and interviews. They are the reason for turning what started as a modest experiment in my academic career to an inspiring and, hopefully, long-term project. Enormous credit is owed to P.J. Chackochan and K.M. George. Not only have they granted me access to the farmers' cooperative, they have also created a welcoming space for ongoing and inspiring dialogue and have shared their precious knowledge throughout the writing of this book. Their willingness to give their time so generously is greatly appreciated. I am also indebted to Anith Puthiyankath and Jörg Volkmann for providing invaluable support with the organization of the data collection, for giving me access and insight into the workings of small food businesses and for our many valuable debates on issues concerning food, sustainability and communication. I am particularly grateful to Jennifer Smith Maguire for being a mentor and critical friend from the very early stages of this project, and to Kelly Touri for offering much-needed moral support during the challenging phases of this work. I also wish to acknowledge the academic community in the Participatory Communication Research Section of the International Association for Media and Communication Research (IAMCR) for the helpful conversations and critical feedback. Last but not least, a special mention must go to the University of Leicester and the School of Media, Communication and Sociology for believing in this work and for providing financial support and a home for conducting the research. The work presented in this book has received funding from the British Academy Small Research Grants and from the University of Leicester College Research Fund and Impact Development Fund.

Contents

List of Images

CHAPTER 1

Introduction

Sustainable development is development that meets the needs of the present, without compromising the ability of future generations to meet their own needs.
The Brundtland Report, 1987

It would be no exaggeration to describe the above quote as the most widely cited definition of sustainable development. Equally, sustainability and the 2030 Agenda for Sustainable Development are amongst the most widely discussed and reported issues as the world grapples with the consequences of climate change, poverty, conflict, and resource scarcity. Sustainable development and the Sustainable Development Goals (SDGs) might seem like a clear and straightforward agenda for ending poverty, protecting the planet, and promoting peace and prosperity. Yet, the truth could not be further than that. This is an idea fraught with contradictions and questions about the future of humanity.

Writing in *Pluriverse: A Post-Development Dictionary*, the sequel of *The Development Dictionary* that was published in 1992, Wolfgang Sachs describes the idea of sustainable development as the expression of an unequivocal mind shift in the geopolitics of development compared to the post-colonial and post–Cold War era from a discourse of progress to that of survival (Sachs 1992, 2018: xiii). Yet, sustainable development is still underpinned by a narrow and one-dimensional perception of

M. Touri, *Sustainable Development and Communication in Global Food Networks*, https://doi.org/10.1007/978-3-030-46119-5_1

development, one that is tied to a market-driven conceptual framework and a mantra that sees appropriate technologies, markets, and institutional policy reforms as the solution (Kothari et al. 2018). As several of the contributors in the Post-Development Dictionary confirm, there is a pressing need for scholarly work to both deconstruct and re-politicise the idea of sustainable development. We need to critique the social structures that underlie ecological destruction and need to bring to the forefront the multidimensional and alternative visions of sustainable development through an interdisciplinary effort. To all the above I would add the need for a critical approach to the idea and measurement of development as a whole, informed not just by expert and scientific knowledge but by the unique knowledge, experience, and voice of the people at the grass-roots level.

It is in this effort that this book aims to contribute first by advancing the idea that genuine sustainable development is more likely to happen through smaller and self-sustained projects rather than institutional efforts that rely on large aid packages and a 'silver bullet' mind set; and second by bringing into focus the critical role of communication, voice, and the social relations that are at the heart of such projects. At a time when achieving genuine sustainable development on a global scale entails a multi-stakeholder approach with resources and knowledge drawn from different social groups, communication could not be more crucial (Servaes and Malikhao 2016). In fact, as the world is faced with an unprecedented humanitarian and economic crisis caused by the COVID-19 pandemic that is unravelling at the time of writing, the need and value of communication and social relations has become even more pressing. The physical distancing policies implemented globally to control the disease have only strengthened the importance of and need for communication and social relations to support and bolster our economies and societies. According to the Chief Executive of the Bank of England Andy Haldane, the crisis has exposed the need for recognising the importance of this social capital in the way we measure the success of our societies, placing well-being at the centre (Haldane 2020). The same should apply to our understanding and measurement of sustainable development, with more attention placed on the voices of the people on the ground.

Sustainable development emerged as a prominent development paradigm in the 1980s, with the year 1987 becoming a landmark for the 'sustainability revolution' with the publication of 'The Brundtland Report' by the World Commission on Environment and Development (WCED). In this, sustainable development was defined as 'development which meets the needs of the present without compromising the ability of future generations to meet their own needs' (Elliott 1994: 4), with core issues including development, energy, food security, and biodiversity. It can be argued that the Brundtland Report was also a landmark for the establishment of what today is called the 'Sustainable Development Goals' (SDGs). The SDGs summarise the agenda agreed by the United Nations' member states to achieve sustainable development between 2015 and 2030, and which continues the Millennium Development Goals (MDGs) that were adopted in 2000. Sustainable development has also been theorised as a form of development and social change, 'a social process, which has as its ultimate objective sustainable development at distinct levels of society' (Servaes and Malikhao 2016: 174).

There has been no shortage of criticism when it comes to the reactions to the UN Sustainable Development Goals, succinctly reflected in suggested alternative readings of the SDGs acronym, such as 'Sustainable Survival Goals' (Sachs 2018: xiii), 'Stupid Development Goals' (The Economist 2015), and 'Senseless, Dreamy, Garbled' (Easterly 2015). The reported inadequacy of the SDG Agenda to offer a realistic development policy has recently received considerable criticism from scholars and experts alike. SDGs have been accused to be driven by a neoliberal ideology, failing to invest in health, correct trade injustices, and protect low and middle-income countries, with the COVID-19 pandemic deepening existing undercurrents even further (van de Pas 2020). Early sustainability debates were initially influenced by *The Limits to Growth* report by the Club of Rome (Meadows et al. 1972), which raised concerns about mass production technologies and consumption patterns and the limitations of perpetual growth. These concerns were captured by the term 'eco-development' and the rejection of the idea that economic growth will spread from the few to the masses; but were ultimately rejected by powerful political actors hence paving the way for sustainable development as the new guiding policy (Gómez-Baggethun 2018: 71).

Although at the time of the COVID-19 pandemic, the UN's focus turned onto global environmental threats and the advancement of SDGs directly linked to climate action, land, and water, the sustainable development discourse has shifted the emphasis from social justice to poverty alleviation in line with the economic imperatives of growth (Gómez-Baggethun 2018). The Rio+20 final declaration advocates the need for economic growth in more than 20 articles, championing the conviction that economic growth and technological progress will offer the solution to environmental issues (Demaria and Kothari 2017). Although the importance of economic growth for human development cannot be denied, it is its non-uniqueness, as a means to development, that needs to be recognised. This requires a shift away from dominant discourses that frame development in line with Western-Northern hegemony and growth, and a recognition of the multidimensionality of human development and well-being. The following section presents three theoretical approaches that exemplify this idea.

Development Beyond Economic Growth: Deconstructing Sustainable Development

The limitations of the economic aspect of sustainable development have been in the public debates at least since the 1990s during which time we also saw the launch of the Human Development Index (HDI) and a shift in the emphasis of development economics from national income accounting to human-centred policies. In the UNDP report *Sustainable Human Development: Concepts and Priorities* (1994), authors Sudhir Anand and Amartya Sen put forward the idea of human development as a goal in itself that safeguards sustainable development by enhancing people's capability to live worthwhile lives, without denying the importance of poverty alleviation for environmental protection. These debates also highlighted the need for deconstructing and reviewing the concept of development as a linear process that relies on material and financial growth, and recognising that well-being and quality of life depend as much on social relations and the environment. According to Demaria and Kothari (2017), deconstructing development can pave the way for a matrix of alternative, richer, and more complex views that challenge the dominant ideas of sustainable development and promote ethical values such as solidarity, interconnectedness, inclusiveness, to name but a few. There are at least three theoretical frameworks that have contributed to this deconstruction and will be presented here. These are Amartya Sen's Capability Approach (1999), Alsop and Heinsohn' Empowerment Framework (2005), and Dorothea Kleine's Choice Framework (2013). All three frameworks contribute to a deconstruction of the idea of sustainable development in a way that reveals the complexity of well-being and the multiplicity of resources, processes, and relations that make up development and can also lead to sustainability.

The Capability Approach

The capability approach can be broadly described as a normative framework, an approach, or a mode of thinking about normative issues that can be used for the evaluation of individual well-being, the design of policies, and proposals about social change in society (Robeyns 2005). What distinguishes this approach from other theories is its emphasis on the need to evaluate well-being and human development not just in monetary terms but also based on different aspects of human life. In other words, human development is approached as multidimensional and should not be reduced to economic growth and income generation. The framework emerged from the work of Nobel Laureate Amartya Sen in 1980s and has contributed decisively to the development of the Human Development Index (HDI) as an alternative indicator to the GDP per capita that shifted the debate from an economic growth focus to one on human development (Kleine 2013: 21). The approach has also been developed by the philosopher Martha Nussbaum (2011) as well as other scholars. Although Sen and Nussbaum disagree on certain features of the theory, they are aligned in their critique of utilitarian approaches to human development. Moreover, as Sen's approach has been described as less prescriptive than Nussbaum's, I will be referring to his work mostly.

In Sen's approach, humans are described as a collection of evolving capabilities, with human development being a multidimensional process that depends on objective (i.e. economic growth) as well as subjective values (i.e. what a person has reason to value). With freedom as its central principle, Sen's theory understands development as a process of expanding the real freedoms that people enjoy (Sen 1999: 3). These substantive freedoms are referred to as 'capabilities', that is, what people are able to be and to do; the range of which make up the 'functionings' available to them (Clark 2005). More specifically, functionings are what an individual 'has reason to value' (Sen 1999: 74). These include valuable activities that contribute to people's well-being, and while some may be focused on survival and material well-being, others emphasise empowerment and self-expression. Based on the concept of functionings, the idea of capabilities encompasses the various combinations of functionings, or the freedom and opportunity to achieve these functionings. Hence, functionings reflect what a person achieves, whereas capabilities correspond to the alternative combinations of functionings that are feasible for a person to achieve in a given environment (Sen 1999: 74–76). From Sen's perspective, the capability approach recognises the value of the economy and economic growth for people's capacity and freedom to expand their capabilities, but without limiting development to just economic growth. Economic growth is seen more like an enabling factor, rather than the only indicator for measuring improvement (Asiedu 2011). According to Nussbaum, human well-being should not be differentiated from agency since well-being is encompassed in whatever an individual values being able to be or do (2011: 200). The human has no higher purpose than empowering herself/himself.

The capability approach is not without its limitations. One of the key questions concerns individuals' responsibility towards the community. If as Sen contends, people 'must take responsibility for doing things or not doing things' in accordance with their agency (1999: 190), it remains unclear how an individual's choices impact those surrounding him. In other words, little attention is given to 'obligation' (Deneulin 2011; Johnstone 2007). Along similar lines, Soper asserts that the capability approach fails to see humans as 'responsible agents assuming accountability to the world beyond their immediate personal concerns' (2004: 112). In other words, and as Dubois and Mahieu (2003) explain, Sen relegates responsibility to second place at best, after liberty, while relationships with other humans are conceived primarily as instrumental rather than as intrinsic to being human. However, if one's capability set ought to grow infinitely, it requires that one person's gain in power does not affect other people. It needs to be mentioned though that the capability approach takes the normative approach of ethical individualism, implying that what ultimately matters is what happens to every individual in the society.

Another central question that emerges from this approach is the extent to which a person has the resources required to achieve certain functionings and capabilities, as well as the ability to use these resources. Here, the specific and subjective circumstances in which people live and exercise choices are key, and they are captured through the notion of conversion factors, that is, the degree to which a person can convert a resource into a valuable functioning (Frediani 2010). The notion of conversion factors draws attention to the impact that the socio-cultural context where people live has on their capacity to aspire and develop functionings and activities he/she values to pursue. The capability approach acknowledges that people may have different opinions about what good life really is, but it also draws attention to the importance of finding a balance between people's freedom to choose the lives they want to lead and their right to have the required knowledge and resources that will enable them to reflect on what they value, set goals, and pursue these goals (Oosterlaken 2015: 5).

Although one cannot ignore these limitations, the capability approach raises important questions about development and empowerment as it moves beyond economic indicators as the measure of improvement. Instead, development is approached as a process of expanding the real freedoms that people enjoy and the things they are able to do. Through elements such as functionings, capabilities, and agency, it draws attention to the need for a deeper understanding of the different aspects of empowerment that underpin human development.

The Empowerment Framework

One of the key contributions of the capability approach stems from the attention that Sen pays to the link between material well-being and psychological and social aspects of development. This connection has been further developed in Alsop and Heinsohn's (2005) and Alsop et al.'s (2006) Empowerment Framework through the concept of asset-based agency and the role of psychological and social assets in increasing people's capacity to make effective choices. In this framework, 'agency is defined as an actor's or group's ability to make purposeful choices' (Alsop et al. 2006: 10), with the achievement of choice constituting a form of empowerment. Amongst the most important assets in understanding and tracking empowerment are psychological and social assets or resources, which might be more difficult to measure than material and financial assets, but they are an indispensable aspect of people's well-being. Psychological resources may include sociability, self-confidence, and capacity to envision change, while social resources include networks and relationships.

For individuals to be able to convert their resources into agency and empowerment, an opportunity structure is also required. According to Alsop and Heinsohn (2005: 9),

> An actor's opportunity structure is shaped by the presence and operation of formal and informal institutions, or rules of the game. These include the laws, regulatory frameworks, and norms governing people's behavior. The presence and operation of the formal and informal laws, regulations, norms, and customs determine whether individuals and groups have access to assets, and whether these people can use the assets to achieve desired outcomes.

However, the authors also recognise the interdependency of resources and opportunity structures in terms of people's need to acquire certain assets to deal with and change existing institutions and norms that impede their agency and development. A useful example refers to women in countries like India and Ethiopia, for whom a change in their psychological resources and their capacity to aspire to a different relationship with their husbands is a prerequisite to changing certain social institutions that govern husband-wife relationships. Although such psychological assets are often ignored, they remain critical in people's well-being and in our understanding of development as a more holistic condition that is not confined to utilitarian measurements.

The Choice Framework

Dorothea Kleine (2013) has drawn on both the capability approach and the Empowerment Framework to develop her Choice Framework to assess the role of information and communication technologies in development. Kleine employs a people-centred approach to development that starts from the individual, is focused on choice, and attempts to map and understand the different elements of the development process as a system. The Choice Framework builds on the idea of empowerment being connected to the availability and use of choice, as outlined by the other two approaches; and Kleine goes a step further in adding the 'sense of choice' and the knowledge that choice exists as another dimension of empowerment (Kleine 2013: 44 in Tufte 2017: 120). The role of non-material assets, such as psychological, social, and informational resources, is also critical in this framework, while Kleine draws attention to the cultural embeddedness of these resources and their interrelation with specific social and cultural contexts. More importantly, Kleine's work brings to the forefront the aspect of 'irrationality' in the process of development, highlighting the 'complex ways in which individuals' emotions, beliefs, conscious and subconscious fears ... shape their decisions and behaviors' (Kleine 2013: 25). As such, she makes a strong case about the need to understand and appreciate the importance of non-economic outcomes that are part of people's well-being and sense of development, which is also demonstrated in the case of Chile. Although Kleine's Choice Framework and its application to the Chilean communities focuses primarily on development outcomes from their engagement with specific information and communication technologies, her work provides valuable empirical evidence of the multidimensional and subjective nature of well-being and the diverse and complex ways in which people make sense of their lives.

The three theories presented here share a common interest in and emphasis on the need to study and understand development through a people-centred approach and based on the complex ways in which people experience development and empowerment within specific cultural, social, and political structures. Despite having underpinned public development debates for decades, the humancentric approach is yet to become a prominent dimension in current sustainable development concepts, theories, and discourses. This book presents an attempt to bring the humancentric dimensions of sustainable development into focus through an interdisciplinary intervention that draws from the global food economy and communication and brings together insights from communication and development studies, economic geography, and agri-food studies. However, the core argument the book aims to advance is not one of rejection of neoliberalism and free market capitalism, but of the possibility of alternative approaches to development where capitalist market relations can coexist with and contribute to non-monetary human development. In order to make these alternative understandings visible and comprehensible, I argue that a new 'language' is also needed, one that shifts our focus away from established and dominant conceptions of sustainable development, capitalism, and the economy, and allows us to look for less universal and more local and alternative explanations. In this book, this new language is found in weak theory, which has been inspired by queer and feminist studies (Chap. 2).

Food: A Mediator in Development and Social Change

> Given its material and social importance, food beckons as an entry point to questions of sustainability that involve joining environmental and socio-economic concerns. (Hinrichs 2010: 19)

Food can be a window to human cultures and a lens for understanding how people connect with each other, with places, and with the natural environment (Steel 2008). At the same time, food production and consumption can reflect the social, economic, and environmental pressing issues that humanity is currently facing. As the food economy continues to globalise and demand for products such as dairy and meat increases, the food system continues to create challenges for natural resources putting immense pressure on the environment and people's well-being (e.g. Kloppenburg et al. 1996; Anderson 2008). Once again, commenting on the impact of the COVID-19 pandemic on the global food economy is inevitable, especially as the global outbreak has brought to light contradictory predictions regarding the future of the global food supply chains. On the one hand, the crisis has sparked a rise in demand for locally grown food, it has intensified calls for self-sufficiency and resilient local food systems and fuelled predictions for the collapse of global food trade. In rich and poor countries alike, local efforts to provide food supplies to local communities have flourished with new pop-up distribution centres and direct-to-home models, while the breakdown of global supply chains has caused a shift to local produce and people are becoming accustomed to managing with fewer and local ingredients (Fennell 2020).

On the other hand, there is also evidence of the resilience of the global food supply chain, with connectivity and technological advancement having helped counteract the impact of the pandemic through a rerouting of supply chains and swapping food supply sources (Huan-Niemi 2020). Given the reliance of most of the Western world on imported food, it is imperative for governments to keep the global food trade alive but also to secure a level of transparency, traceability, and accountability that will safeguard people's health and well-being. It also means reducing industry concentration and protectionism, protecting biodiversity, and supporting developing countries (Economist 2020; Institute for European Environmental Policy 2020). As the Chief Economist of the Food and Agriculture Organisation, Maximo Torero Cullen, said recently, global food trade needs to keep going, especially considering it provides a significant source of income and welfare to low- and middle-income countries (FAO 2020). This level of reliance renders global food trade ever more crucial for the sustainability of farming communities in countries of the Global South, but it also makes the need for a more human-centred, innovative, and coordinated approach to sustainable development more pressing (Institute for European Environmental Policy 2020). It is specifically the humancentric element of the global food market that this book wishes to bring to focus, in the pursuit of a more nuanced understanding of sustainable development.

This book approaches the global food economy through the prism of the so-called alternative food production and distribution practices that are captured by the umbrella term 'Alternative Food Networks' (AFNs)

(Forssell and Lankoski 2015: 63). Alternative Food Networks will be discussed in Chap. 2. At this point, suffice it to say that the contribution of AFNs in the dominant sustainability discourses and in the conception and development of the UN's SDGs is primarily framed around environmental considerations, food security, and hunger eradication (see FAO 2019), and around the expectation that sustainable food systems should become more environmentally sound and economically viable for a larger percentage of community members (Feenstra 2002: 100 in Hinrichs 2010: 20). Their sustainability promises are premised on the adaptation and promotion of environmentally benign farming practices, such as organic farming, value addition from differentiated and artisan production methods, as well as on the reduction of 'food miles' (Renting et al. 2003; Marsden et al. 2000; Schonhart et al. 2009). The reduced physical distance that certain types of AFNs are associated with, such as Community Supported Agriculture (CSA) and farmers' markets, has been at the heart of localisation arguments that champion localism and local-scale food networks as inherently more beneficial for ecological sustainability, food security, and social justice than globally extended networks. Such arguments are predicated on the assumption that local food systems of production and consumption will yield economic, environmental, and social benefits for communities, while the notion of 'community' is also defined in terms of locality, ignoring the fact that communities can exist at all scales (Born and Purcell 2006). Underlying these discussions is the idea of the conflation of global agriculture with capitalist food systems. In this case, local networks are championed as spaces of resistance against the growing industrialisation of the food system (Kaufman 2004, 338).

Such arguments and dualisms have already been contested, with scholars advocating for recognition of the significance of interdependencies between local and global food networks for sustainable development (Goodman et al. 2014; Kneafsey et al. 2008). This book contributes to this body of work by adding a communication perspective and unearthing the critical role played by relations and communication practices in global food networks and why these matter for sustainable development. Food production and consumption enfold multiple socio-material processes and vibrant social worlds that create opportunities for unpredictable social change and development and that need to be considered and understood better (Preiss et al., 2017; Ofstehage 2017). More importantly, food is produced through an array of human relations and social networks that are crucial for sustainable development and that also provide an entry to questions and considerations of the multifaceted nature of development, and particularly, the human and social side of it. The case of globally extended Alternative Food Networks encapsulates this relational nature of food production-consumption and therefore lends itself to deeper investigations of the role of food production and consumption in sustainable development. In her recent book *Voice and Participation in Global Food Politics*, Alana Mann describes food as the mediator in a process of social learning for social change and for re-establishing connections with what we eat and with the people who produce it (Mann 2019: 3). To this I would add that food is also the mediator in communication practices and social relations that are essential to sustainable development.

Global Alternative Food Networks are also a manifestation of the global market and capitalist relations and a component of the global food supply chain architecture. Although they are permeated by neoliberal discourses of economic development and they epitomise many of the ills of the global market, they also rely on trusted relations between food growers, intermediaries, and consumers that exhibit a more social and humancentric side of market relations. For this reason, they offer a platform where an alternative side to capitalism can be made visible; and once again communication provides a valuable tool that allows us to revisit traditionally economic actors, such as businesses and consumers, as agents of sustainable development.

Finally, the case of global AFNs inevitably raises questions about the cooperation and relationships between producer and consumer countries that are traditionally located in the Global South and North, respectively. Sustainable development discourses promoted through academia, international organisations, and the media often encourage an understanding of these relations as relations of dependency that obscure the heterogeneity of the South and undermine the agency of its people. Ultimately, deconstructing sustainable development requires that the voices of these people be heard. Communication is the thread that can bind together the different dimensions of sustainable development as this is experienced by the people. In fact, Sen makes an explicit reference to the role of communication in human freedom, quality of life, and formation of values (Sen 2009: 335–337). In this book, communication is also approached as a tool that can bind together different practices and social relations that underpin the global food economy towards sustainable development.

Communication in Sustainable Development

The focus on communication encompasses the effort of this book to advance an understanding of development that is determined by people and their relations, where communication becomes 'the essential thread that binds them together' (Fraser and Villet 1994). A focus on communication not as a deliberate intervention but as a process that permeates social life and social change can help unearth the less visible routes and paths through which people's knowledge and wisdom can inform theories and practices of development. Communication, interpersonal and mediated, is at the heart of sustainable development with institutions such as UNESCO and the World Bank's Communication for Sustainable Development in Operations (SDO) unit working towards strengthening the capacity of institutions and individuals to make the best use of available communication resources for their sustainability. For communication to contribute to sustainable development, it needs to encourage a shared understanding of communities' values and everyday problems and promote solutions that are desired by the communities above all. Yet, at the same time, the complexity and ambivalence of sustainable development make communication difficult to achieve and be effective (Kleinen-von Königslöw 2018). According to a PANOS report, one of the reasons why the promises made in the Millennium Development Goals (MDGs) failed was the lack of recognition by policymakers and development experts of the vital role of communication (Wilson et al. 2007). The report recognises that communication and sustainable development are inseparable as it is through better communication and information processes among people and between people and governments that communities can achieve change. This requires investment in building healthy civil society, good governance, and media and communication systems that will facilitate the participation of marginalised communities in social, political, and economic processes and change. Although communication remains vital and integral in sustainable development, and it is through communication, transparency, and knowledge sharing that global risks can be addressed, it is not explicitly mentioned in any of the 17 SDGs (Servaes 2013).

In the field of communication studies, there is now a rich depository of theories and methods that employ communication to address specific sustainable development issues, including rural development, poverty reduction, natural resource management, and gender equality. A combination of methods including communication campaigns, marketing techniques, as well as more creative and participatory methods such as storytelling and participatory video are often used to raise awareness and influence attitudes on environmental issues (Servaes and Lie 2015). Academic perspectives on communication and sustainable development can be broadly organised in two categories: science communication and development communication. While the former engages mostly with environmental and climate change communication, and with more strategic approaches to communication, the latter pays more attention to participatory communication approaches for rural and human development. This book is positioned in the latter category, and the role of communication in sustainable development is examined through the prism of the field of academic and applied research that is commonly known as Communication for Development and Social Change, which will be introduced in Chap. 3. Interestingly, and despite the field's consistent preoccupation with sustainability and development issues, it remains a separate and distinct field and is rarely positioned within the academic and research area of sustainable development. Nevertheless, it has provided a wealth of empirical and conceptual knowledge of the power of communication, as an intervention and as an everyday practice, to affect economic and social change and to give people tools to make change and achieve appropriate and sustainable development (Fraser and Villet 1994). It is the aim of this book to bring to light the value that this field can offer to our understanding of sustainable development not as the outcome of unidirectional and market-driven development processes but as an experience and an embodied subjectivity that occurs and is shared through social relations. The analysis of communication is complemented with theories and concepts from the broader field of communication studies, which are fruitful in demonstrating the different types of communication and relations that happen outside institutionalised development programmes and can still contribute to community well-being and sustainable development.

These relations will be explored through the context of global food networks and the interaction/communication between three actors: farmers, small businesses, and consumers across three continents. This communication can be interpersonal/direct or mediated/indirect. In either case, the focus is on communication that is informal, 'silent', and less visible but has the power to lead to social change in significant ways. The power of this communication is encapsulated in the concept of *weak communication* that will be introduced in Chap. 3. It aims to capture the integral role of communication in development through communication practices and relations that are embedded in the structures of the global food market and include trade relations (Chap. 4), representation (Chap. 5), and information and knowledge exchange through the spaces of the internet (Chap. 6). The concept of weak communication also embodies the performativity of knowledge and the need for a new language that will pave the way for the disruption of dominant discourses of sustainable development. In this book a new language is found in queer theory and feminism scholarship and will be unfolded in Chaps. 2 and 3 setting the context for the remaining chapters. In Chap. 4, the idea of weak communication is explored through the trade relations that connect farmers in India with small businesses in Europe and North America. These relations are based on a close interaction that can prove empowering for the farmers' community, contributing to different aspects of their social, economic, and environmental sustainability. Chapter 5 engages with communication in the form of farmers' public representation through the marketing and promotional material employed by fair trade organisations and brands. Here, the concept of weak communication illuminates the 'weakly' theorised impact of these representations on the farmers' well-being. It also paves the way for a novel empirical investigation of the empowering effect that the process of self-representation can have on farmers' well-being and sustainability. An experiment that involves Indian farmers' production of their own videos and representations confirms the need to lift the voices of grassroots and marginalised communities in the study of sustainable development. In Chap. 6, the focus shifts on another weakly theorised aspect of the relations and communication practices that underpin the global food market, namely the interaction between producers and consumers across the globe. Apart from food, people's consumption involves information and knowledge about food, which shapes their habits and behaviour not just in their private sphere but also towards the environment and the distant farmers. The chapter demonstrates the power that a more direct and personal interaction between producers and consumers can have on consumers' perceptions and attitudes towards food and, more importantly, their perceptions about their own role and responsibility towards the environment and the people who grow our food. In this case, farmers' expertise and digital technologies become essential instruments in consumers' role in sustainable development. Through the relations and communication practices that are unpacked in these three chapters, the book aims to contribute to the research and practices of sustainable development in a two-fold manner. On the one hand, it seeks to highlight the need for communication to be prioritised by and placed centrally in the work of scholars in the social and natural science disciplines. On the other hand, it aims to draw the attention of communication scholars and practitioners to the development opportunities that reside in communication practices that happen organically, such as in the structures of the global food market. It is by engaging with such communication practices further that the field of communication can contribute to development discourses beyond communication studies.

The Case of Kerala in South India

The core ideas of this book are explored and illustrated through the experience of a farming community in the district of Wayanad in the Indian state of Kerala. Agriculture is the most important sector of the Indian economy, accounting for 18 per cent of India's gross domestic product (GDP) and for more than 50 per cent of the population's employment (IBEF 2019). India is also among the countries whose agriculture and food economy trajectory has been shaped by land reforms and development programmes promoted by international development organisations and scientists (Chap. 2). By focusing on the state of Kerala, this book offers empirical and experiential knowledge that is rather un-representative of India. Instead, the Kerala case provides a suitable and rich context for challenging conventional understandings of development and illustrating the central role that food production plays in this.

The case of Kerala's development experience represents what is often described as the 'development model' thanks to the state's significant success in attaining a high HDI and high standard of living without rapid economic growth. This achievement both contradicts and challenges the presumptions and predictions of all economic development theories that see social development, demographic transition, and human well-being as positively correlated to GDP (El Khoury 2016; Thottathil 2014). Kerala's success, on which Amartya Sen's capability approach has also drawn on, is predicated on augmenting basic human capabilities through a long history of reforms that dates to the nineteenth century (Parayil and Sreekumar 2003). The trajectory of Kerala's development success is beyond the scope of this book, but it is worth recognising some of the factors that contributed to this success, achieved with little foreign aid and with a key role played by grassroots movements and community economies (El Khoury

2016). Among those factors are the meaningful land reforms, access to primary and preventative healthcare, promotion of education and literacy, especially among women, efforts to reduce social and income inequalities, and public action and mobilisation of peasant farmers and agricultural and industrial workers (Parayil and Sreekumar 2003). A particularly significant factor was the open and cosmopolitan outlook of Kerala that resulted from its rich history of trade relations and cultural links with the rest of the world, especially as a spice exporter, since as early as 3000 BCE (Chattopadhyay and Franke 2006). Although Vasco da Gama's arrival at Calicut paved the way for the domination of the Portuguese over the spice trade and for Western colonialism in the sixteenth century, this era also established Kerala's full integration into Western capitalism and the global food economy (Sreekumar 1993). Currently, Kerala continues to play a key role in the country's export business with more than 80 per cent of the state's agricultural products being exported (El Khoury 2016: 303).

But while trade has been one of the key drivers in Kerala's development, it also proved to be the source of adversity for the state, especially following trade liberalisation and WTO policies that led to a sharp decline in the prices of agricultural products. This was coupled with high price instability leading to loss of income and high unemployment rates (Joseph and Joseph 2005; Shiva 2001; Vasavi 1999). Kerala's traditional exports became subjected to the volatility of global markets, and the situation deteriorated further in 2001 with the removal of quantitative restrictions (QRs) on 714 mostly agricultural imports, the lowering of tariff levels, and the fierce competition for exports in the international market (Jeromi 2007; Joseph and Joseph 2005; Palackal 2019). During the 1980s and 1990s, the Indian government's fiscal crisis meant that along with the decline in agricultural output, Kerala's manufacturing sector also slowed (Thottathil 2014).

Kerala's agrarian crisis was inevitable given the state's dependence on the world market (Mohanakumar and Sharma 2006). At the same time, the crisis encouraged an ecological countermovement. This was to battle the damaging consequences of the practices dictated by industrial and commoditised agriculture, such as mechanization, mono-cropping, and excessive use of chemical fertilisers, that transformed agriculture into a cash crop–based business (Palackal 2019; Thottathil 2014). The countermovement was instigated by the Kerala government, and in a policy document released by the Biodiversity Board and Kerala's Agriculture Department in 2010, the government introduced a strategy for the promotion of organic farming and conservation of agro-biodiversity across the entire state. The organic farming programme—accompanied by the required certification schemers—has provided new opportunities for farmers to work with distant buyers and participate in globalised commodity chains on more favourable terms, while it has also enabled the formation of local groups and cooperatives (Thottathil 2014: 108–109). Although the state has clearly been a key actor in this initiative, the role of organic farmers from the periphery of Kerala's agribusiness has been decisive in the development of this countermovement and the establishment of a collective identity not just among farmers but also NGOs and other like-minded individuals (Palackal 2019).

The significance of this organic countermovement is two-fold: on the one hand, it signals Kerala's embracement of traditional, natural farming methods as a path to a more sustainable future. On the other hand, it is indicative of its vibrant civic culture and its long tradition of cooperative movements and community economies that have been an integral part of Kerala's economy (El Khoury 2016: 256). From this perspective, Kerala represents a social, cultural, and agricultural context where movements, such as the organic countermovement, articulate the significance of the community and cooperative culture that underpins the formal economy. It is a context that lends itself to the study of the social relations that underline formal economic transactions in the global food economy and through which alternative understandings of development can be drawn.

Situated in the north east of Kerala and home to Kerala's organic export movement and organic certification institutions (Thottathil 2014: 85), the Wayanad district is set high on the Western Ghats, which is a UNESCO World Heritage Site and a biodiversity hot-spot. Wayanad has a long history of trade activity as evidence suggests the existence of trade routes to Wayanad even prior to the colonisation of India (Cheriyan 1999). It is also believed that the hill tribes were involved in international trade activities, trading in forest products (Münster 2012). Agriculture is the backbone of Wayanad's economy, and although the area is emerging as a tourism destination, it is also considered a backward district characterised by land alienation and degradation, with very few agro-based industries. It is currently included in the Backward Regions Grant Fund Programme (Kjosavik and Shanmugaratnam 2015; Martin and Rajesh 2013).

Wayanad is also known for being one of the most seriously distressed agrarian districts in India and one of the most affected by farmer suicides, having been hit by a series of economic and ecological crises since 1990s

(Münster 2012; Vakkayil 2010). As Münster (2012) explains, the chemicals that were heavily used to boost productivity and profit during the economic boom years resulted in a catastrophic damage to the soil and a rise in cancer incidents among the local population. A severe drought in 2004 and an epidemic of plant diseases made matters even worse. The ecological crisis coincided with the economic crisis and crash of market prices for cash crops that hit Kerala during the late 1990s, leaving many farmers with heavy debts (Vakkayil 2010). For many of them, suicide was the only way out of this crisis, given the lack of support from the state and the minimal government investments (Krishnakumar 2004). Although the economic and ecological crisis remain the first reason for the majority of suicide cases until the mid-2000s, farmers' suicides in Wayanad have also been attributed to other reasons, including physical illness, family problems, and even the pressures of a consumer culture (Jeromi 2007). After 2000, Wayanad witnessed the rise of a consumer culture that put additional pressure on families to borrow money in order to keep up with a rising middle class, but which also culminated in further debt and suicides (Münster 2012).

The suicides and agricultural hardships led to the involvement of the Catholic Church with a group of religious activists forming the Indian Farmers' Movement (INFAM) in 2000. Amongst the key priorities of the movement was the replacement of chemical-intensive practices with organic farming methods, the development of alternative marketing channels, and the writing-off of small farmers' debts, a demand that was actually met by the Indian government in 2008. INFAM also realised the importance of organic certification for the opening of new market channels in Europe and America, which would give the farmers' products a competitive edge in foreign markets. This led to the establishment of 'Organic Wayanad',[1] a charitable society and cooperative that serves as an internal control system and ensures farmers meet the criteria for organic certification. It provides training and support during the farmers' organic conversion and liaises with foreign buyers. The formation of Organic Wayanad was followed by the setting up of the Indian Organic Farmers' Producer Company (IOFPCL), a producer company that is a combination of a private limited company and a cooperative society and acts as a liaison entity between the organic farmers and the buyers (Thottathil 2014; Vakkayil 2010). Apart from the significant enhancement to the soil

[1] http://vanamoolika.org/organic-wayanad

ecology, the benefits of organic farming methods have encouraged many farmers in Wayanad to focus on organic products for export, turning Wayanad's agrarian sector into export oriented (Thottathil 2014).

Kerala and the district of Wayanad represent a rich socio-cultural context that lends itself well to examining and understanding the various possible ways in which food, communication, and well-being come together in the context of food networks and the global food economy. The economic development of Kerala and the gradual spread of consumption culture are juxtaposed with many farmers' search for a more spiritual connection to nature and the land through more radical alternative farming practices, such as zero budget natural farming, especially following the struggles they faced during the agrarian crisis (see Münster 2016, 2017). This context raises interesting questions about the different ways in which farmers and their communities negotiate well-being and sustainable development. Moreover, the entrepreneurial spirit and high literacy levels of many farmers have given them the confidence to engage in export relations and negotiations with foreign buyers.

The farmers who participated in the research presented in this book are all members of the 'Organic Wayanad' Cooperative. Between the start of this research in 2014 and at the time of writing, the number of members registered with the cooperative has doubled, from around 200 to 420 members, all of whom had organic certification for the products they were growing for export and for the domestic market. From those, half of them were also fair trade. The size of the farms varied, and it ranged between 12 acres and a quarter of an acre. The areas that were visited for the research and where the farmers were located and interviewed were in the town of Pulpally and the surrounding villages, which included Thrissilery, Kundala, Thondarnad, Seetha Mount, and Amarakuni.

A Note on Gender

Although gender power relations have not been at the heart of this book, a note on the role of women in the agricultural sector of Kerala, and specifically in the villages of Wayanad where the data has been collected from, is paramount. A trend that is worth noting and was also present in the farming community that took part in this research concerns the fact that women's presence has been more visible in their role as housewives rather than as farmers. This is despite their unquestionable contribution to farming activities that range from cultivation to marketing (Kodoth 2004).

One explanation is found in the changing cropping pattern and the growing size of the farms. It has been reported that the move from subsistence farming to cash crops has driven more women into domestic work, while women in households with larger farms are also less involved in farming (Arun 2012). At the same time, household dynamics and women's headship mean that women continue to be involved in the maintenance of the farms and in economic decisions that concern the household and the farm. This is especially when their husbands take office jobs that keep them away for long hours (Arun 2012). In Wayanad, women work mostly as co-farmers and their critical role in the running of the farms was acknowledged by their husbands. However, during the fieldwork, very few women agreed to take part in the discussions, which is a typical challenge when conducting research in patriarchal societies (Momsen 2006). Without dismissing or downplaying the significance of women's knowledge and experience, their underrepresentation in this project should be seen in the context of the broader limitations of any study that focuses on a narrow case of investigation, especially in complex local landscapes such as India. Hence, in this study, women's knowledge is underrepresented in a similar way that the knowledge and experience of other farming communities and tribes has also been excluded. The empirical evidence presented in this book is by no means generalisable, and the aim of this work was to offer initial insights in sustainable development through a communication perspective rather than draw conclusive inferences.

Research Process, Positionality, and Reflexivity

The methodological tools employed in this book fall under what is best described as a transdisciplinary, mixed-method approach that draws from the field of social sciences. It aims to bring to light some of the lessons that can be learnt from communities and communication processes that so far have received little attention in the study and understanding of sustainable development. Although the departing point of the work that has brought this book to fruition was a straightforward qualitative inquiry into buyer-farmer communication in trade relations (Chap. 4), the data that emerged from this inquiry and my immersion as a researcher in the work of the specific Indian community of farmers not only generated new knowledge around the role of communication, it also sparked a process of reflexivity among all the parties involved—farmers, buyers, and researcher—a joint process of knowledge production and new insights for the researcher and the participants. This changed the nature of the research to more interactive, collaborative, and participatory as the voice of the participants became embedded in the design of the research process. Hence, without negating or undermining the existence of a certain degree of researcher bias, this is mitigated by the valuable knowledge and experience that participants brought into this process. At the same time, it is this collaborative and participatory nature that turned this research from an attempt to observe and understand relations and communication processes to a project that facilitated social and economic change in the farming community that took part in this project. In this case, the participatory and community-based approach enabled the connection of different scales in the food network, from the farming community to the local society and the global market.

The organic way the research developed allows for a chronological description of the different methodological steps that were taken for the completion of this project. Hence, the point of departure was the inquiry into the interpersonal communication that is embedded in trade relations and connects farmers with the foreign buyers that facilitate the farmers' entry into the global food market. The emphasis on interpersonal communication as a form of communication that supports the structures of food networks but remains invisible was driven by the need to create a pathway to a more nuanced understanding of the multifaceted nature of sustainable development through a people-centred approach and through the voices and the perceptions of the people that were most affected, that is, the farming communities. This part of the inquiry consisted of focus groups with farmers in Wayanad and five of their buyers in Europe and North America. The choice of a qualitative research method was the most appropriate for achieving a more holistic understanding of the communication process and for capturing the underlying meanings and issues that matter to the participants.

The first focus groups with the farmers took place in February 2014 and involved 30 participants. They were combined with individual interviews with the Chairman of IOFPCL at the time and the Coordinator of Organic Wayanad. The interview process was facilitated by a local interpreter. The buyers were interviewed between August 2014 and January 2015. Four of the interviews were conducted on Skype and one in person in the UK. Four of those buyers are small-medium enterprises, wholesalers, and retailers, while one of them is a bigger alternative trading organisation. At the time of the interviews, all buyers had organic certification, and four of them were also certified by fair trade.

The focus groups and interviews aimed to shed light on their relationships and communication beyond the strictly market exchanges, as well as the possible ways in which they contribute to the farmers' growth and development. The findings from these group and individual discussions (Chap. 4) not only revealed the importance of interpersonal communication for farmers' agency, but they also paved the way for the continuation of this research as both farmers and buyers expressed the desire to build closer relations with each other, while farmers also signalled their need to be heard and valued by the local society and the global community of consumers. It was from this point onward that the research assumed a participatory nature with the conversion of the findings from the first round of focus groups into an action plan that involved the farmers, two of the buyers that had already been interviewed and two new buyers. The plan comprised a storytelling experiment and the production of short videos (Chap. 5). The idea was conceived in September 2016 during a second round of focus groups with the farmers in Wayanad. The initial objective was to revisit the farmers, observe possible changes in the participants' perceptions and practices that could be linked back to their relation and communication with the buyers, and collect new data from more farmers. The fact that as a researcher I had already built a relationship with the farmers and the farmers had also began to think more collectively about their communication with the buyers, with consumers, and with each other worked as an enabling factor for this part of the research. Hence, the focus groups offered farmers a safe setting and 'communicative' space to not just start a conversation with each other (Bergold and Thomas 2012) but to also develop new ideas and a plan of action.

It was during this stage of the process where the participants also began to take more ownership of the research (Russo 2012) by deciding to produce a number of short videos with the use of a mobile phone camera, or short written stories and still images. During these discussions action-oriented goals were also set, with the farmers planning dissemination events for sharing the videos with the local community. The story creation took place between October and December 2016, leading to twenty videos and six written stories. In January 2017, a second block of five focus groups were held with 20 of the farmers who took part in the videos to discuss and reflect on their experience of making the videos. The participation of the buyers occurred in a rather natural way, as four buyers (all in Europe) that work more closely with the farmers' group became aware of the storytelling experiment and they offered to organise screening events for their own consumers in their respective locations. In research terms, the buyers' involvement in this stage of the research was the result of the knowledge that was co-produced between the researcher and the farmers. At the same time, the screening events organised by the buyers produced new insights that informed the farmers' actions and the researcher's analysis and evaluation.

Following the completion of the video production, the videos and stories were shared with food consumers in the UK through a website that was designed using a free web hosting provider, and where all material was uploaded. Focus groups with consumers were held in the UK in May and June 2017. Ten mix-gender focus groups were held in total, consisting of five participants each. Five of the focus groups took place in London and five in the Midlands. The groups were homogenous in terms of age, while all participants had expressed some interest in organic and fair trade products. Although the sample is not representative, the objective of these group interviews was to gain initial insight into consumers' reactions to the farmers' self-representation, rather than to generalise the findings to the entire population. The participants were asked to watch the videos and/or read the stories of the farmers prior to the discussion.[2] Finally, a third block of six focus groups was conducted in Wayanad in September 2017, with 20 farmers who had taken part in the videos and 20 farmers who had not taken part, but had been informed about the project through their colleagues. During these discussions, the consumers' responses were delivered to the farmers through a combination of verbal narration, a presentation of quotes and extracts from the discussions, and a presentation of short videos that were shot during the focus groups with the consumers.

The focus groups with the farmers who followed from the completion of the video production and from the consumers' reactions offered a space for discussion and reflection for the farmers who ultimately informed their action. As they reflected on the consumers' enthusiastic feedback (Chap. 5) and on the value of the videos for their own internal

[2] The participants in the UK were recruited through a recruitment agency. The moderation was done jointly by the author and an independent professional moderator.

use, two new plans of action emerged. One plan included the organisation of screening events in two different regions, in Kerala and Karnataka, which took place in September and October 2017. The second plan concerned the production of new videos that would focus on the different stages of production of specific spices, such as turmeric, to be used for marketing purposes, through the foreign buyers, and for raising awareness among local farmers for the benefits of natural organic farming. Since then, I have held occasional individual interviews with farmers and officers from Organic Wayanad to discuss their experiences and aspirations.

In light of this book's quest to encourage an understanding of sustainable development through the prism of communication and the voices of the farmers in Kerala, I feel that it is my obligation to recognise my positionality and the fact that my education, and my background and experiences as a European person would inevitably affect my observations and analysis. However, although I am an 'outsider' to the social, cultural, and political context of the farming community that I studied, this is partly mitigated by the close friendships and personal relations that I had developed with people from Kerala long before this study took place. These friendships were critical not only in facilitating the collaboration with the participants, the conduct of the interviews, and the translation/interpretation of the discussion, but also in enabling me to interact with the farmers even outside the data gathering spaces (e.g. in community events) and forge relations with them. Although interactions are an expected outcome of conducting fieldwork with qualitative methods (Patton 2002), my previous contact and knowledge with the local culture and the presence of Keralite friends during the fieldwork gave me an identification and connection with the farmers that was crucial for gaining their trust and acceptance. The trusted relationships also proved significant for maintaining reflexivity through prolonged engagement and repeated interviews with the same farmers as well as member checking. At the same time, although my familiarity with the community's culture enhances the research process, being less familiar with their work and their farming practices enabled me to maintain a healthy distance, placing the farmers in the position of the expert (Berger and Malkinson 2000).

References

Alsop, R., & Heinsohn, N. (2005). *Measuring empowerment in practice: Structuring analysis and framing indicators* (World Bank Policy Research Working Paper 3510). https://doi.org/10.1596/1813-9450-3510.

Alsop, R., Bertelsen, M. F., & Holland, J. (2006). *Empowerment in practice: From analysis to implementation* (The World Bank, number 6980). World Bank Publications.

Anand, S., & Sen, A. (1994). Sustainable human development: Concepts and priorities. In S. Anand & A. Sen (Eds.), *Human development index: Methodology and measurement* (Human Development Report Office, Occasional Papers). Available at https://ora.ox.ac.uk/objects/uuid:98d15918-dca9-4df1-8653-60df6d0289dd/download_file?file_format=application/pdf&safe_filename=HDI_methodology.pdf&type_of_work=Report. Accessed 2 Mar 2019.

Anderson, M. D. (2008). Rights-based food systems and the goals of food system reform. *Agriculture and Human Values, 25*(4), 593–608.

Arun, S. (2012). 'We are farmers too': Agrarian change and gendered livelihoods in Kerala, South India. *Journal of Gender Studies, 21*(3), 271–284.

Asiedu, C. (2011). Information communication technologies for gender and development. *Information, Communication & Society, 15*(8), 1–31.

Berger, R., & Malkinson, R. (2000). "Therapeutizing" research: The positive impact of family-focused research on participants. *Smith College Studies in Social Work, 70*(2), 307–314. https://doi.org/10.1080/00377310009517594.

Bergold, J., & Thomas, S. (2012). Participatory research methods: A methodological approach in motion. *Historical Social Research, 37*(4), 191–222. https://doi.org/10.2307/41756482.

Born, B., & Purcell, M. (2006). Avoiding the local trap. *Journal of Planning Education and Research, 26*(2), 195–200.

Chattopadhyay, S., & Franke, R. W. (2006). *Striving for sustainability: Environmental stress and democratic initiatives in Kerala*. New Delhi: Concept Pub. Co.

Cheriyan, P. J. (1999). *Perspectives on Kerala history*. Trivandrum: Cultural Publication Department.

Clark, D. A. (2005). *The capability approach: Its development, critiques and recent advances* (ESRC Global Poverty Research Group working paper series 32). Manchester: Institute for Development Policy and Management. http://www.gprg.org/pubs/workingpapers/pdfs/gprg-wps-032.pdf

Demaria, F., & Kothari, A. (2017). The post-development dictionary agenda: Paths to the pluriverse. *Third World Quarterly, 38*(12), 2588–2599. https://doi.org/10.1080/01436597.2017.1350821.

Deneulin, S. (2011). The capability approach: A note on two interpretations. Paper presented at a Conference on Human Development, University of Bielefeld, Germany, 25–26 June 2011.

Dubois, J. L., & Mahieu, F. R. (2003). Personalism, capability and sustainability. From Emmanuel Mounier to Emmanuel Levinas: Ethics for a socially sustainable development. Paper presented at the *3rd International Conference on the Capability Approach, 7–9 September 2003, Pavia*. http://ethique.perso.sfr.fr/pavie2.htm

Easterly, W. (2015, September 28). The SDGs should stand for senseless, dreamy, garbled. *Foreign Policy*. http://foreignpolicy.com/2015/09/28/the-sdgs-are-utopian-and-worthless-mdgs-development-rise-of-the-rest/. Accessed 10 Oct 2017.

Economist. (2015). The 169 commandments, 26 March, 2015. Available at https://www.economist.com/leaders/2015/03/26/the-169-commandments. Accessed 16 July 2019.

Economist. (2020). How to feed the planet: The global food supply chain is passing a severe test. Available at https://www.economist.com/leaders/2020/05/09/the-global-food-supply-chain-is-passing-a-severe-test. Accessed 10 May 2020.

El Khoury, A. (2016). *Globalization development and social justice: A propositional political approach*. Oxon/New York: Routledge.

Elliott, J. (1994). *An introduction to sustainable development*. London: Routledge.

FAO. (2019). *The state of food security and nutrition in the world: safeguarding against economic slowdowns and downturns*. Rome: Food and Agriculture Organization of the United Nations. Available at http://www.fao.org/3/ca5162en/ca5162en.pdf. Accessed 15 Feb 2020.

FAO. (2020). *A battle plan for ensuring global food supplies during the COVID-19 crisis*. Available at http://www.fao.org/news/story/en/item/1268059/icode/. Accessed 21 June 2020.

Feenstra, G. (2002). Creating spaces for sustainable food systems: Lessons from the field. *Agriculture and Human Values, 19*(2), 99–106. https://doi.org/10.1023/A:1016095421310.

Fennell, S. (2020). *Opinion: Local food solutions during the coronavirus crisis could have lasting benefits*. Available at https://phys.org/news/2020-04-opinion-local-food-solutions-coronavirus.html?fbclid=IwAR0bPUz0rLGUFa0HGXho7bMxsysJmZwh-pFU0Nbh06BEsfsFB3f2I7ek9oc. Accessed 1 May 2020.

Forssell, S., & Lankoski, L. (2015). The sustainability promise of alternative food networks: An examination through "alternative" characteristics. *Agriculture and Human Values, 32*(1), 63–75.

Fraser, C., & Villet, J. (1994). *Communication: A key to human development*. Rome: Food and Agriculture Organization of the United Nations. Available at http://www.fao.org/3/t1815e/t1815e00.htm. Accessed 10 Apr 2019.

Frediani, A. A. (2010). Sen's capability approach as a framework to the practice of development. *Development in Practice, 20*(2), 173–187. https://doi.org/10.1080/09614520903564181.

Gómez-Baggethun, E. (2018). Sustainable development. In A. Kothari, A. Salleh, A. Escobar, F. Demaria, & A. Acosta (Eds.), *Pluriverse: A post-development dictionary* (pp. 71–73). New Delhi: Tulika Books.

Goodman, D., DuPuis, M. E., & Goodman, M. K. (2014). *Alternative food networks: Knowledge, practice and politics.* New York: Routledge.

Haldane, A. (2020). *Responding to COVID-19: Social capital and the social sector in the 21st century.* Available at https://www.probonoeconomics.com/news/responding-covid-19-social-capital-and-social-sector-21st-century. Accessed 20 June 2020.

Hinrichs, C. C. (2010). Conceptualizing and creating sustainable food systems: How interdisciplinarity can help. In A. Blay-Palmer (Ed.), *Imagining sustainable food systems: Theory and practice* (pp. 17–36). Farnham, Surrey: Ashgate.

Huan-Niemi, E. (2020). *The global food supply chains are capable of providing enough food to feed the world population, but the rich world should help the poor over the COVID-19 crisis.* Available at https://www.luke.fi/en/blog/the-global-food-supply-chains-are-resilient/. Accessed 1 July 2020.

IBEF. (2019). *Indian economy growth rate and statistics.* Available at https://www.ibef.org/economy/indian-economy-overview

Jeromi, P. D. (2007). Farmers' indebtedness and suicides: Impact of agricultural trade liberalisation in Kerala. *Economic and Political Weekly, 42*(31), 3241–3247.

Johnstone, J. (2007). Technology as empowerment: A capability approach to computer ethics. *Ethics and Information Technology, 9*(1), 73–87.

Joseph, B., & Joseph, K. J. (2005). Commercial agriculture in Kerala after the WTO, *6*(1), 37–57. https://doi.org/10.1177/139156140500600103

Kaufman, J. (2004). Introduction to special issue on planning for community food systems. *Journal of Planning Education and Research, 23*(4), 335–340.

Kjosavik, D. J., & Shanmugaratnam, N. (2015). *Political economy of development in India: Indigeneity in transition in the state of Kerala.* London: Routledge.

Kleine, D. (2013). *Technologies of choice? ICTs, development, and the capabilities approach.* Cambridge, MA: MIT Press.

Kleinen-von Königslöw, K. (2018). *The role of communication for sustainable development and the SDGs: Objectives and specific challenges, presentation.* Available at: https://www.sd-network.eu/pdf/doc_workshops/2018%20berlin/presentations/Katharina%20Kleinen%20von%20Koenigsloew.pdf. Accessed 2 Sep 2019.

Kloppenburg, J., Hendrickson, J., & Stevenson, G. W. (1996). Coming into the Foodshed. *Agriculture and Human Values, 13*(3), 33–42. https://doi.org/10.1007/BF01538225.

Kneafsey, M., Holloway, L., Venn, L., Dowler, E., Cox, R., & Tuomainen, H. (2008). *Reconnecting consumers, producers and food: Exploring alternatives.* London: Berg.

Kodoth, P. (2004). Gender, property rights and responsibility for farming in Kerala. *Economic and, Political Weekly, 39*(19), 1911–1920.

Kothari, A., Salleh, A., Escobar, A., Demaria, F., & Acosta, A. (2018). Introduction: Finding pluriversal paths. In A. Kothari, A. Salleh, A. Escobar, F. Demaria, & A. Acosta (Eds.), *Pluriverse: A post-development dictionary* (pp. xxi–xl). New Delhi: Tulika Books.

Krishnakumar, R. (2004). Driven to death. *Frontline, 21*(9). Available at https://frontline.thehindu.com/the-nation/article30222292.ece. Accessed 21 Jan 2020.

Mann, A. (2019). *Voice and participation in global food politics.* Abingdon/New York: Routledge.

Marsden, T., Flynn, A., & Harrison, M. (2000). *Consuming interests: The social provision of food.* London: UCL Press.

Martin, P., & Rajesh, K. (2013). Critical evaluation of backward region grant fund in Kerala. *Asian Journal of Research in Social Sciences and Humanities, 3*(4), 56–67.

Meadows, D., Meadows, D. L., Randers, J., & Behrens, W. W., III. (1972). *The limits to growth: A report for the club of Rome's project on the predicament of mankind.* New York: Universe Books.

Mohanakumar, S., & Sharma, R. K. (2006). Analysis of farmer suicides in Kerala. *Economic and Political Weekly, 41*(16), 1553–1558.

Momsen, J. H. (2006). Women, men and fieldwork: Gender relations and power structures. In V. Desai & R. Potter (Eds.), *Doing development research* (pp. 44–51). London: SAGE.

Münster, D. (2012). Farmers' suicides and the state in India: Conceptual and ethnographic notes from Wayanad, Kerala. *Contributions to Indian Sociology, 46*(1), 181–208. https://doi.org/10.1177/006996671104600208.

Münster, D. (2016). Agro-ecological double movements? Zero budget natural farming and alternative agricultures after the neoliberal crisis in Kerala. In B. B. Mohanty (Ed.), *Critical perspectives on agrarian transition: India in the global debate* (pp. 222–244). New Delhi: Routledge.

Münster, D. (2017). Frontier in ruins: Internal colonization and the making of an agricultural landscape in South India. In P. Zemisch, U. Münster, J. M. Zickgraf, & C. Lang (Eds.), *Soziale ästhetik, atmosphäre, medialität: beiträge aus der ethnologie* (pp. 97–107). Münster: Lit.

Nussbaum, M. (2011). *Creating capabilities: The human development approach.* Cambridge, MA: The Belknap Press of Harvard University Press.

Ofstehage, A. (2017). Encounters with the Brazilian soybean boom: Transnational farmers and the Cerrado. In P. Pingali & G. Feder (Eds.), *Agriculture and rural development in a globalizing world: Challenges and opportunities* (pp. 60–72). London/New York: Routledge.

Oosterlaken, I. (2015). *Technology and human development*. New York: Routledge.

Palackal, A. (2019). Organic agriculture in Kerala: A counterdiscourse from the margins. *Sociological Bulletin, 68*(2), 169–182. https://doi.org/10.1177/0038022919848254.

Parayil, G., & Sreekumar, T. T. (2003). Kerala's experience of development and change. *Journal of Contemporary Asia, 33*(4), 465–492. https://doi.org/10.1080/00472330380000291.

Patton, M. (2002). *Qualitative research & evaluation methods* (3rd ed.). Thousand Oaks, CA: SAGE.

Preiss, P., Charao-Marques, F., & Wiskerke, J. S. C. (2017). Fostering sustainable urban-rural linkages through local food supply: A transnational analysis of collaborative food alliances. *Sustainability, 9*(7), 1155. https://doi.org/10.3390/su9071155.

Renting, H., Marsden, T., & Banks, J. (2003). Understanding alternative food networks: Exploring the role of short food supply chains in rural development. *Environment and Planning A, 35*, 393–411.

Robeyns, I. (2005). The capability approach: A theoretical survey. *Journal of Human Development, 6*(1), 93–117.

Russo, J. (2012). Survivor-controlled research: A new foundation for thinking about psychiatry and mental health. *Forum: Qualitative Social Research, 13*(1), Art. 8.

Sachs, W. (Ed.). (1992). *The development dictionary: A guide to knowledge as power*. London: Zed Books.

Sachs, W. (2018). The development dictionary revisited. In A. Kothari, A. Salleh, A. Escobar, F. Demaria, & A. Acosta (Eds.), *Pluriverse: A post-development dictionary* (pp. xi–xvi). New Delhi: Tulika Books.

Schonhart, M., Penker, M., & Schmid, E. (2009). Sustainable local food production and consumption: Challenges for implementation and research. *Outlook on Agriculture, 38*(2), 175–182. https://doi.org/10.5367/000000009788632313.

Sen, A. K. (1999). *Development as freedom*. Oxford: Oxford University Press.

Sen, A. K. (2009). *The idea of justice*. Cambridge: Cambridge University Press.

Servaes, J. (2013). Introduction: The kaleidoscope of text and context in communication. In J. Servaes (Ed.), *Sustainability, participation and culture in communication: Theory and praxis* (pp. 1–24). Bristol: Intellect.

Servaes, J., & Lie, R. (2015). New challenges for communication for sustainable development and social change: A review essay. *Journal of Multicultural Discourses, 10*(1), 124–148. https://doi.org/10.1080/17447143.2014.982655.

Servaes, J., & Malikhao, P. (2016). The role and place of communication for sustainable social change (CSSC). *International Social Science Journal, 65*(217–218), 171–183.

Shiva, V. (2001). Recovering biodiversity. *Social Change, 31*(1–2), 21–37. https://doi.org/10.1177/004908570103100204.

Soper, K. (2004). Rethinking the "good life": The consumer as citizen. *Capitalism Nature Socialism, 15*(3), 111–116.

Sreekumar, T. T. (1993). *Urban process in Kerala: 1900–1981.* Thiruvananthapuram: Centre for Development Studies, St. Joseph's Press.

Steel, C. (2008). *Hungry City: How food shapes our lives.* Chicago: Chatto & Windus.

Thottathil, S. E. (2014). *India's organic farming revolution: What it means for our global food system.* Iowa City: University of Iowa Press.

Tufte, T. (2017). *Communication and social change: A citizen perspective.* Cambridge: Polity.

United Nations. (1987). *Our common future*, The Brundtland Report (Report of the World Commission on Environment and Development). Available at https://www.are.admin.ch/are/en/home/sustainable-development/international-cooperation/2030agenda/un-_-milestones-in-sustainable-development/1987%2D%2Dbrundtland-report.html. Accessed 25 May 2019.

van de Pas, R. (2020). *The Coronavirus pandemic and the irrelevance of the SDGs.* Ghent Centre for Global Studies. Available at https://www.globalstudies.ugent.be/the-coronavirus-pandemic-and-the-irrelevance-of-the-sdgs/. Accessed 20 July 2020.

Vakkayil, J. D. (2010). INFAM in Wayanad, Kerala. In M. Harper (Ed.), *Inclusive value chains: A pathway out of poverty.* Singapore: World Scientific.

Vasavi, A. R. (1999). Agrarian distress in Bidar: Market, state and suicides. *Economic and Political Weekly, 34*(32), 2263–2268.

Watts, D., & Kettunen, M. (2020). *Global sustainable development in the aftermath of the COVID-19 pandemic.* Institute for European Environmental Policy. Available at https://ieep.eu/news/global-sustainable-development-in-the-aftermath-of-the-covid-19-pandemic. Accessed 20 June 2020.

Wilson, M., Warnock, K., & Shoemaker, E. (2007). *At the heart of change: The role of communication in sustainable development.* London: PANOS Institute.

CHAPTER 2

Development in Global Food Networks: Uncovering the Sociability of the Global Food Economy

Introduction

This chapter sets the broader context of this book first by establishing the connection between development, food production, and the global food economy. It begins with a review of historical key events that have shaped the global food system, including the restructuring of agriculture during colonial rule, the Green Revolution—a controversial technocratic model driven by a top-down technology and knowledge transfer from the North to the South—and the promotion of international trade and export-oriented agriculture as a path to development. Such occurrences were an early manifestation of the economistic tendency that has characterised development initiatives, especially in the Global South, focusing on profit maximisation at the detriment of social life and the environment, and contributing to the emergence of a clear econocentric discourse of development. The review paves the way for the introduction of globalised Alternative Food Networks (AFNs) that emerged as a form of resistance and support for development in the South.

Alternative Food Networks (AFNs) are a prominent case of resistance movements and can be loosely divided in two broad categories: local networks that emphasise the need for shorter supply chains and territorialised value chains; and globalised networks such as the organic and fair trade movement. It is the latter category that this chapter is concerned with, focusing specifically on the development opportunities generated through

M. Touri, *Sustainable Development and Communication in Global Food Networks*, https://doi.org/10.1007/978-3-030-46119-5_2

the participation of Southern farmers in these networks. The developmental dimension of both organic and fair trade networks has long been recognised by international development agencies such as the World Bank and the Department for International Development (DFID), as well as by European policymakers. Organic agriculture is perceived to play a key role not just in promoting environmental integration and sustainable development but also in providing local and national development benefits and supporting the livelihoods of farming communities (Egelyng and Høgh-Jensen 2006). Along similar lines, investments from international development institutions and the European Commission into entrepreneurship opportunities associated with fair trade highlight the movement's role in serving the aims of the development industry (Dolan 2005).

Despite representing a form of resistance to the unequal power relations and inequalities that underpin global food trade, the case of AFNs has also raised concerns about weaving together environmental protection and economic expansion, and reinforcing the same econocentric discourse of sustainable development they are meant to resist. More importantly, although such tendencies have been widely critiqued by agri-food and development scholars alike, existing critiques risk reproducing the language and dominance of hegemonic neoliberalism, creating an impasse on conceptions of development in the global food market.

The chapter goes on to argue that in order to unearth the grassroots and more alternative practices and conceptions of sustainable development, we need a new line of thinking and a new language in the way we talk about and understand the workings of the global food economy and the different shapes that development can take. We need a language that can provide space for the mindsets of marginalised groups and alternative disciplines. In this chapter, a new line of thinking and speaking is found in the theory of diverse economies by feminist economic geographers Katherine Gibson and Julie Graham. This is a framework that was inspired by queer theorist Eve Sedgwick's 'weak theory', a theory that invites social scientists to embrace surprise and alternative theoretical investigations, and it has begun to inform the theorisation and empirical examination of AFNs. Drawing on the key ideas from these approaches, the chapter introduces an alternative way of speaking and thinking about development, social change, and the global food economy, with the aim to reveal the dynamic nature of the economic and social relations that constitute food production, and the diversity of development opportunities that reside in the global food market.

Agriculture, Food, and Development in the Global Economy: A Neoliberal Discourse of Rural Development

> For a colonised people, the most essential value, because the most concrete, is first and foremost the land: the land which will bring them bread and, above all, dignity. (Fanon 1982: 44)

As the world becomes increasingly focused on sustainable futures and the achievement of the internationally agreed Sustainable Development Goals, food production and agricultural progress become ever more critical for development. Agricultural and food systems have become a powerful lever in accelerating overall economic growth and modernisation, particularly in countries in the Global South. Agricultural performance has helped transform many emerging economies through the industrialisation of processing, reorganisation of distribution, and commoditisation of food (Caron et al. 2008; Pingali and Feder 2017). At the same time, it has created major challenges for countries to remain competitive in a growing and globally integrated food economy. Such challenges include not just environmentally related risks but also human rights issues, concerns around dependency on imported food and unsuitable technologies, concentration of food processing and of distribution chains, market instability, and reduced access to credit (HLPE 2011; IBRD/ World Bank 2007). These challenges are the legacy of a long history of agricultural practices and reforms that have culminated in the power imbalances that characterise today's global food economy, and the emergence of resistance movements that seek to curtail the damaging globalising economic forces of modern food.

The trajectory of the global food economy has been inextricably linked to European colonialism, which has not only shaped food production and distribution in decisive ways but has also given rise to prevailing discourses of the food economy's role in rural development. One of the most dramatic consequences of colonial rule was the restructuring of agriculture and the conversion of farming land into large plantations for the production of export commodities such as sugar, which resulted in the colonies' dependence on export. Colonialism also shaped the social relations that underpinned food production creating divides between elites and small farmers and deepening social and ethnic inequalities. These were further exacerbated during the land reforms—redistribution, restructuring, nationalisation—that took place after decolonisation (Weis 2007: 90–99). For instance, in Haiti smallholders and agricultural labourers were exploited by a minority elite that maintained cash crop estates (Trouillot 1994), in China private land ownership was eliminated (Weis 2007), while in India, land reforms proved successful only in the states of Kerala and West Bengal. The rest of the country experienced a modernisation reform implemented by the government of Jawaharlal Nehru, which led to the development of a slightly larger class of entrepreneurial commercial farmers, in place of the small rural aristocracy, and which did nothing to address issues of land inequality, poverty, and food insecurity (George 1990: 184). The struggles that many countries in Africa, Asia, and Latin America faced in dealing with poverty and food insecurity inevitably paved the way for the Green Revolution, a so-called technological 'fix' that was promoted by international development organisations and scientists as an alternative reform for addressing problems of famine, food shortages, and dependence on foreign aid (Weis 2007).

The Green Revolution initiated in Mexico with financial support from the Rockefeller Foundation and the USDA. It consisted of the introduction of higher yielding new varieties of hybrid plant seeds, fertilisers, agrochemicals, irrigation systems, and agricultural technologies that were developed in laboratories in the Global North. This was an industrial system of agriculture that aimed at achieving more efficient use of natural resources, increasing productivity, and maximising yields (Thottathil 2014). This process would then lead to industrialisation through freeing up and/or pushing out labour from agriculture into the new factory system and producing more agricultural produce from less land and less human labour (Perkins 1997). Industrialisation of agriculture was aligned with the turn towards the more commercialised farming that was enforced during colonial rule. As a result, subsistence farming that served local consumption was gradually abandoned as farmers turned to the cultivation of monocrops and cash crops for export (Thottathil 2014; Weis 2007).

In many respects, the consequences of the Green Revolution proved devastating. According to data from FAO, more than 70 per cent of the world's agrobiodiversity that was originally maintained in smallholder agroecosystems was lost from farming. As agriculture became more mechanised, homogenised, and chemical-based, farmer communities grew smaller and less autonomous, eventually losing control over their markets. Smallholder farmers had to rely on government subsidies and loans to buy seeds and fertilisers, with many of them facing bankruptcy when their land became sterile (see Hewitt de Alcántara 1976; Shiva 1991; Weis 2007). The Green Revolution was a success in terms of an increase in the aggregate supply of grain, but a failure in terms of putting an end to hunger (Perkins 1997: 258). For critics such as the Indian physicist Vandana Shiva (1991), these consequences were unsurprising. The diffusion of agricultural innovations from capitalist countries was based on the interests of US-British capitalism and the powerful elites in the bourgeoisie national governments. The introduction of miracle seeds, fertilisers, and agricultural technologies into the Global South was an opportunity to create a new market for the large US and British manufacturers. These global corporation interventions were promoted as a pathway to solving the food problem in the poorer countries and containing the peasant unrest that was aimed at regaining the peasants' lost land (Dutta 2011; Shiva 1991).

The development initiatives implemented during colonialism and the Green Revolution paved the way for the aggressive promotion of a neoliberal model of development by international organisations and national governments. We begin to see this in the way the international agricultural trade changed especially after WWII. Agricultural production and food export from the South to the North became the primary strategy for industrialisation, through the profits, foreign exchange earnings, and taxes generated from trade that were then channelled to industrial expansion. The process included the introduction of additional advanced technology for plantations and the incorporation of more farmers into market relations (Bernstein 2000). This required the countries' gradual dependence on cheap food imports from the North, which were seen as a crucial strategy to stimulate economic growth and boost national development by bringing down the cost of living and, subsequently, the cost of labour too (Weis 2007).

The neoliberal norms that underpinned these strategies were further enhanced through structural adjustment programmes (SAPs) and recommendations enforced by the International Monetary Fund (IMF) and the World Bank, the liberalisation of agricultural trade, and reduced domestic support and protectionism for countries in the Global South. Structural adjustment programs were imposed on many countries as a measure to avoid bankruptcy when the governments of said countries had to default on foreign loans following deteriorating export prices and increased costs of oil imports that created stark trade imbalances between the North and the South. With structural adjustment programs in place, and as free trade became the dominant economic paradigm, governments in the Global

South were forced to divert their agricultural sectors to export production and maximise export earnings in order to repay debts (Lyon and Moberg 2010). These were measures that inevitably increased farmers' dependence on monocultures and few mostly unprocessed crops, and their vulnerability to price crashes and crop failures. In several cases, farmers were put out of business as with the rice farmers in Haiti after the lift of import restrictions on subsidised rice from United States (see Oosterveer and Sonnenfeld 2014).

Dependence on the international agricultural trade 'locked' farmers in the South into an unequal relationship with the North and with the international market more broadly, placing them in the often impossible position of having to continuously improve their capacity to produce the quality that is expected by foreign markets, but lacking the necessary institutional support and equipment to do so (Oosterveer and Sonnenfeld 2014). Also, the industrialisation model of rural development and growth that was promoted by institutions and governments meant that small subsistence farmers would suffer the most, being forced to consolidate into large farms or be replaced by large food-selling farmers. Inherent in this development logic was the misperception among policymakers about the contribution of small farms to economic growth. This misperception assumes that small farms are inefficient, cannot be commercial or evolutionary, an assumption that several studies have disproved (see Lipton 2017; Murray McIntire 2017).

The era of colonialism and the Green Revolution represent landmarks in the production and establishment of a powerful econocentric discourse that promoted the politics of scientific legitimacy and market expansion as the answer to poverty eradication and development. This is a discourse that is predicated on a top-down and linear technology and scientific knowledge transfer from the centre to the periphery. This ideology is based on the belief that change happens through research, extension, and education, where knowledge and technology is designed by research institutes in North America and Western Europe and then adopted by users in the rest of the world (Biggs 1990; Henderson and Clark 1990; Leeuwis and Aarts 2011). It is a technocratic model driven by a utilitarian rationality and with a clear economistic tendency, aiming to maximise production and profits at the detriment of social life and the environment (Pachón-Ariza 2013). This economistic tendency has long been the target of critiques but continues to drive development programmes.

Although the neoliberal discourse that has shaped the global food economy created impetus for the emergence of varying forms of movements and measures to support development in the South, their effectiveness remains questionable. A case in point is the policy recommendations championed by UN organisations such as the International Fund for Agricultural Development (IFAD), an agency that is dedicated to the eradication of poverty and hunger. IFAD recognises the dangers that an unregulated international food market bears for the farmers when they lack the resources and knowledge. However, IFAD's developmental strategies are characterised by a top-down approach to knowledge dissemination and management and remain linked to neoliberal concepts of development, growth, and competition. Along similar lines, the Food and Agriculture Organization of the United Nations (FAO), an influential research and development agency, advocates strategies that are shaped by the conviction that the road to development is through economic growth and competition and focuses on integrating farmers into the global food economy (see Schanbacher 2010).

The emphasis on hunger reduction, agricultural growth, and increased technology utilisation by organisations such as IFAD and FAO have also informed the discussion around food security as a development framework. Based on FAO's definition,

> Food security exists when all people, at all times, have physical and economic access to sufficient, safe and nutritious food to meet their dietary needs and food preferences for an active and healthy life … it concerns the availability of food, access to food, the biological use of food, and stability [of the other three elements over time]. (FAO 1996)

Food security emerged after WWII as a development framework, discussed and promoted through the 1948 Universal Declaration of Human Rights and the 1966 International Covenant of Economic, Social and Cultural Rights (Wittman 2011). In the context of FAO, food security is perceived as a multidimensional concept that supports poverty reduction through the fight against hunger and the need for balanced nutrition and has informed relevant legislative reforms that focus on the availability, access to food, and food utilisation. Achieving food security necessitates a multifaceted strategy that should combine the integration of farmers in the global economy, as well as the commitment of national governments and international institutions to growth, trade, and social and economic policies to ensure access of the poor to food resources. For FAO, the global economy is full of opportunities for food security and development that can only materialise through the training of farmers in using new agricultural techniques and technologies (Gordillo and Jerónimo 2013; Schanbacher 2010). Food security is underpinned by a human rights approach that has been integrated into policies to battle hunger. The emphasis on the right to food was first established in the Universal Declaration of Human Rights (UDHR), and was later integrated into the International Covenant on Economic, Social and Cultural Rights (ICESCR) (Gordillo and Gómez 2005). More recently, the food security discourse has incorporated two central goals: that food production needs to increase by 50 per cent by 2030 and that food production needs to double by 2050 to feed a world population of 9 billion (House of Commons 2009: 9).

Yet, the remit of food security has been questioned, especially regarding the ways in which food and hunger is framed as a problem and the solutions proposed to it. For instance, FAO's approach to food and hunger as a problem of insufficient trade and distribution overshadows deeper problems with systems of production and consumption (McMichael 2005; Patel 2009). Such frames have inevitably shaped many of the food policies that end up focusing on aid and technological development to increase global food production as a solution (Wittman 2011). According to Schanbacher (2010), there is an economic conception inherent in food security, as approached by international organisations, where food production is understood and promoted as the sum of economic relations among economic agents that interact through competition and self-interest. This is also reflected in IFAD's recommendation for food security through the transition from self-sustainable food production to larger economic relations and farmers' entry into competitive markets for agricultural inputs. In essence, food security reinforces the principles of the Green Revolution and aligns with the Western model of rural development by promoting a perception of rural development primarily as economic growth that can be achieved through increased agricultural production, an idea that has transcended the public policies in many countries of the Global South (Gutiérrez-Escobar 2018).

The Power of Discourse in the Global Food Economy and Sustainable Development

From the era of colonialism until the introduction and establishment of the food security policies, agricultural development efforts have followed an econocentric discourse. According to post-colonial theorists, this is a discourse that embodies the power of the North to control ideas of economic development in the South through its power to name, represent, and theorise, and the same discourse is inherent in current conceptions of sustainable development (Escobar 2001; McEwan 2008: 165). This is a discourse cultivated through a preoccupation of politicians, development planners, and field administrators with financial aid being given to projects that would generate tangible input and evidence of development and would allow politicians to demonstrate achievements. This preoccupation both influenced and was influenced by the research priorities around industrial crops that were widely marketed and served commercial interests (Chambers 1974). This discourse was also encouraged by a top-down model of education concerned with scientific research conducted in universities by scientists who were 'deeply conditioned to believe that they know more than farmers, that their knowledge is superior, and that they should be the people who determine what research should be done and how it should be conducted' (Chambers and Jiggins 1987: 39). Chambers' work also draws attention to the role of the rich, powerful, and urban-based professionals and their perceptions about rural communities in the establishment of this discourse. In his article, 'Rural Poverty Unperceived: Problems and Remedies' (Chambers 1981), he explains the failure of urban-based professionals to understand the integrated nature of poverty and development, since their understanding is based solely on brief rural visits that prevent them from forming accurate and unbiased perceptions. It is through these misperceptions that the language, words, and concepts that support development discourses are ultimately defined (Chambers 1981).

The obsession with scientific research methodology and refusal to work with and learn from farmers reflected a scientific arrogance that undermined indigenous knowledge in favour of 'Western Expertise' characterised by a lack of common sense. This 'arrogance' was also at the core of the expert knowledge that defined the problem of malnutrition and hunger and the modernisation solutions that found expression in the Green Revolution bypassing the communities' local knowledge and experience.

The current sustainable development discourse is believed to be shaped by the same logic and thinking of economic expansion that fails to reflect a rethinking of the dominant ideologies and of the role of the West in global inequities (Escobar 2001). The role of knowledge generation institutions, such as academia, research centres, and universities, has been pivotal in the cementing of these discourses. The proliferation of development sciences and sub-disciplines has led to the professionalisation and institutionalisation of the discipline on food, nutrition, and sustainable development. It was a gathering at the Massachusetts Institute of Technology (MIT) in 1971, where food and nutrition experts met, that offered the platform for the official launch of the discipline (ibid). This development discourse encapsulated not just an economic, political, and social control of the West over the rest of the world but a discursive control too (Melkote and Steeves 2015: 100).

The economic, political, and discursive control reflected in dominant discourses of sustainable development has been attributed to a deeper political economy problem that concerns the privatisation and liberalisation of agriculture, the slashing of state subsidies, and the shift from sustainable local forms of agriculture to commoditised capitalist-driven models (Dutta 2011). The same political economy problem is also believed to have encouraged the increase in development programmes that legitimise free markets as the road to development and social change and as substitutes for state-driven and state-funded responsibility (Wilkins and Enghel 2013). Without negating the legitimacy of these critiques, the emphasis they place on neoliberalism risks reinforcing an understanding of sustainable development that fails to consider the multiplicity of interpretations that are possible and ends up reinforcing the dominant discourses that are meant to be critiqued. A more nuanced understanding of sustainable development requires a different view and a new way of thinking on the interplay between neoliberalism, capitalism, and development. This can be achieved through an interdisciplinary approach that draws on agri-food studies and economic geography. Alternative Food Networks are an interesting case as they encapsulate the different sides of the interplay between capitalism and development. On one hand, they display the advantages and disadvantages of development in a market context. On the other, they have paved the way for economic geographers to propose a new language and understanding of the global economy based on which new ideas and discourses of sustainable development can also be built.

Globalised Alternative Food Networks

Alternative Food Networks (AFNs) are a movement that represents a critique of industrialised food systems, the neoliberal relations that conventional markets are normally associated with, and the environmentally damaging effects of large-scale agriculture (Goodman et al. 2014). The underlying philosophy of AFNs is about establishing more direct connections between ethically minded consumers who care about the implications of their consumption practices and producers who want to use production practices that counter the dominant market logic (Whatmore and Clark 2006). This philosophy has also been underpinned by the principles of food sovereignty, a critical alternative to food security that emerged in 1996 through La Via Campesina, a transnational social movement of farmer organisations, in an effort to fight the neoliberal model of food production and protect the rights and needs of small farmers. Especially following the 2008 food crisis, governments had to prioritise food policies that focused on the support and protection of small-scale family farming (McMichael 2011). A decisive moment for La Via Campesina was the conference in Tiaxcala, Mexico, where a coalition of farmers' and indigenous people's movements met in 1996 to discuss the implication of the globalised food economy on their communities and ecologies. It was during this conference that the concept and framework of 'food sovereignty' was born (Wittman 2011). An international forum that took place in Nyéléni, Mali in 2007 defined food sovereignty as follows:

> Food sovereignty is the right of peoples to healthy and culturally appropriate food produced through ecologically sound and sustainable methods, and their right to define their own food and agriculture systems. It puts the aspirations and needs of those who produce, distribute, and consume food at the heart of food systems and policies rather than the demands of markets and corporations. It defends the interests and inclusion of the next generation. It offers a strategy to resist and dismantle the current corporate trade and food regime, and directions for food, farming, pastoral and fisheries systems determined by local producers and users. Food sovereignty prioritises local and national economies and markets and empowers peasant and family farmer-driven agriculture, artisanal – fishing, pastoralist-led grazing, and food production, distribution and consumption based on environmental, social and economic sustainability. Food sovereignty promotes transparent trade that guarantees just incomes to all peoples as well as the rights of

> consumers to control their food and nutrition. It ensures that the rights to use and manage lands, territories, waters, seeds, livestock, and biodiversity are in the hands of those of us who produce food. Food sovereignty implies new social relations free of oppression and inequality between men and women, peoples, racial groups, social and economic classes and generations (Declaration of Nyéléni, in Via Campesina 2007).

Food sovereignty is concerned with the implications of the current neoliberal approach to food production, the unfair trading practices of bilateral free trade agreements, and the corporatisation of global food networks. It also champions a shift from the hierarchical and corporate approach to research and development to a more bottom-up approach characterised by more decentralisation of decision-making in the production of social and economic knowledge (Ishii-Eiteman 2009).

Although food sovereignty is often being described as an 'anti-trade' movement, in reality, food sovereignty activists advocate for trade relations that will be favourable to and support the economic, environmental, and social priorities of farming communities across the globe (Wittman 2011). Hence, food sovereignty aims to advance community development, small-scale food production, cooperation, well-being, and sustainable development as opposed to a neoliberal vision of a globally integrated food economy that is based on competition and profit-making (Schanbacher 2010). Food sovereignty strategies seek to dis-embed food from broader global market relations. They aim to highlight economic, environmental, and social concerns around agricultural production and re-embed food production to broader social and cultural demands, including the need for sustainable production (Zerbe 2014: 86). They also promote transparent and just trade and the rights of producers to be in control of and manage their land, seeds, livestock, and biodiversity (Declaration of Nyéléni, in Via Campesina 2007). Many of the core principles of food sovereignty are promoted through AFNs, which are traditionally characterised by relational chains that are based on transparency and traceability and rely on consumers' willingness to pay a premium price in return for products that were grown in an environmentally friendly and ethical manner, and can allow farmers to stay in business (Stamm 2008). In theory at least, AFNs represent alternative economic platforms with the capacity for development and social change through re-building relations of care, trust, and commitment based on closer proximity and mutual understanding (Kneafsey et al. 2008). Since the late 1990s, when social scientists began examining alternative food systems, the scholarship on AFNs has been characterised by a growing interest and debate around questions of alterity and the distinction between 'alternative' and 'conventional' systems of food provisioning (e.g. Goodman et al. 2014; Kneafsey et al. 2008; Maye and Kirwan 2010). The debate has now moved beyond dualistically opposed definitions and towards recognition of the relational contingency of 'alterity' that is understood in terms of the relations that connect the different production-consumption activities (Kneafsey et al. 2008; Maye and Kirwan 2013).

The literature on AFNs is now vast and covers a range of initiatives, including community supported agriculture, organic agriculture, fair trade, and foods with a geographical indication of origin (e.g. Ilbery and Maye 2005; Renting et al. 2003). In a comprehensive review of AFN conceptualisations reflected in the literature, Forssell and Lankoski (2015) identified some of the core characteristics of AFNs, which included environmentally benign production processes, reduced distance between producers and consumers, notions of localness, small size of networks, transparency, and new forms of market governance based on a redistribution of power and the sharing of economic risk and/or resources. In most, if not all, cases, these initiatives are designed to support small farmers by providing an alternative channel to the mainstream market, as well as to promote environmentally sustainable production practices and allow consumers to have access to fresh quality food (Freidberg and Goldstein 2011).

Although AFNs were born as a form of resistance to the industrialisation of food on a global scale, it is important to distinguish between alternative food provisioning practices that were designed to encourage local networks, shorter supply chains, and territorialised value chains, on one hand, and networks that focus on adding transparency and fairness to global supply systems. This distinction is crucial because it permeates the origins of AFNs as a concept and a practice, as well as the different conceptions of development and social change that different models of AFNs embody. Hence, despite the vast scholarly attention that AFNs have received, the literature remains disproportionally focused on Europe and North America mainly because AFNs were born as a form of resistance to the industrialisation of food in the Global North (Millner 2016; Freidberg and Goldstein 2011). For European scholars, alternative networks embody and encompass the development of niche markets where small businesses and farmers in peripheral rural locations can distribute their products and improve their livelihoods (e.g. Marsden et al. 2000, 2002: Marsden and Smith 2005). Academic and policy discourse in Europe focuses on the role of AFNs in linking producers of quality, artisanal, and provenance foods with consumers, enabling producers to add value and creating opportunities for rural regeneration through more diversified farm livelihoods and local income generation (Goodman et al. 2014: 68). This is reflected in what Ilbery and Kneafsey (1998: 330) describe as the 'considerable potential for a relocalization of the agrofood system in which locally produced quality products and services, with authentification of geographical origin and traceability, are transferred to regional and national markets'. The European-focused studies of Van der Ploeg et al. (2000) and Renting et al. (2005) offer valuable insights into the efforts made to stimulate local rural development through attaching products to places. The importance of 'locality' is exemplified through attempts to revive more traditional market relations, such as the phenomenon of direct marketing through box schemes and farmers' markets, as a form of resistance against the corporatisation of food production (Lamine 2005). In the North American school of thought, the emphasis is on environmental and social justice and resistance to the neoliberal relations that sustain a corporate-controlled food supply (Goodman 2004). Local food networks in the United States are an expression of resistance towards corporate farming through farmers' markets and community supported agriculture (CSA).

However, what counts as an AFN in the countries of the Global South remains debatable. For instance, in a study of a pilot organic box scheme in Kenya that was run through a non-governmental body but failed to take off, Freidberg and Goldstein (2011) explain how the implementation of AFN models in the Global South requires in-depth understanding and appreciation of the historical and socioeconomic contexts that inform on-the-ground practices. Abrahams' (2007) study on South Africa shows that local AFNs represent an alternative consumption space for the urban poor and culturally diverse communities, rather than an attempt to shorten the food chain or foster social relations between producers and consumers. Nor is local food necessarily trusted, given the levels of pollution found in the soil and water of places like Nairobi (Thatiah 2005). Along similar lines, the lack of efficient policy and regulation in India's agriculture sector and the abundant subsidies available for fertilisers has resulted in frequent incidents of food contamination from locally produced food (Thottathil 2014). The farmers who were interviewed in Wayanad also flagged up the longer-term dangers posed by the unregulated use of fertilisers. They explained that even if people are not dying instantly, there are instances of ulcer and cancer that have been traced back to food.

Two models of Alternative Food Networks that capture global dynamics of food production and consumption with the participation of producers and consumers across the globe are the organic and fair trade movements. These movements represent organised efforts to boost rural development in the Global South and are designed to protect farmers from the environmental damages of industrialised food production and the unjust trading relations of the global food economy. In this way, the two movements embody an alternative discourse to the global food economy that provides a basis on which to build a new understanding of sustainable development.

Organic Agriculture and Development

The questionable results of the Green Revolution and the persistence of poverty and food insecurity led donors and governments to push for a commercialisation of organic agriculture to allow smallholders to access the global organic market. Although the Green Revolution increased agricultural production, this was accompanied with environmental degradation. The Green Revolution packages of high-yield crop varieties were also unsuitable for many smallholders and areas lacking the necessary infrastructure to support them (Altieri 2009; Bazuin et al. 2011). Moreover, weak local governance, unfavourable policies, and restricted access to credit meant that most of the poorer farmers were unable to benefit from the packages and agricultural technologies that were introduced to them (Pingali 2012). Organic agriculture presented a plausible measure for contributing to the welfare and security of vulnerable farming communities both economically and environmentally (Bakewell-Stone et al. 2008). Environmental degradation is directly linked to poverty and income loss when farmers' livelihoods depend on natural resources. Organic agriculture can help mitigate these challenges by offering a holistic agricultural production system that creates the right conditions for environmental and economic sustainability (Seufert et al. 2012).

Organic agriculture has its origins in the biodynamic agriculture that was inspired by the philosophy of Rudolf Steiner in Germany, during the 1920s, before it was taken up by the youth protest movements in the 1960s and turned into an alternative professional farming method (Oosterveer and Sonnenfeld 2014). Although initially all farming was considered organic, a clear distinction between organic and non-organic farming developed and was based on third-party certification, and verification that farming practices conform to organic standards (Albersmeier et al. 2009). The International Federation of Organic Agriculture Movements (IFOAM) defines organic agriculture as a holistic system that is driven by the principles of health, ecology, fairness, and care for humans, as well as ecosystems, and promotes biological practices that are aligned to biodiversity conservation, such as crop rotation and the use of local manure. Through a combination of tradition, innovation, and science, organic agriculture aims to benefit the environment and promote fair relationships and a good quality of life for all involved (Oosterveer and Sonnenfeld 2014; Seufert et al. 2012). By building up natural resources, organic agriculture can contribute to improving human capacity in farming communities (Parrott and Marsden 2002). Worldwide, the organic agricultural land has increased massively during the last decades, from 11 million hectares in 1999 to more than 35 million hectares (Willer and Kilcher 2009).

Notwithstanding the environmental benefits, it was the economic benefits and profitability of organic farming—through increasing consumer demand from Europe and United States, the premium prices and the lower production costs—that has incentivised a number of governments in the South to overcome regulatory barriers in order to access foreign markets for organic products (Nemes 2009). Hence, participation of smallholders in the organic movement has been approached by donors and national governments as another route for the integration of farming communities in the global food economy through export to US and European markets.

According to official reports, during the first 10 years of the new millennium, worldwide organic food markets expanded by 10–15 per cent, as opposed to a much smaller growth in conventional markets, while European markets in particular grew by 8 billion Euros in the span of 5 years (Willer 2011; UNCTAD 2008). Fresh fruits and vegetables are in high demand due to their alleged nutritional qualities, with supermarkets demanding a year-round supply of seasonal fresh food products (Oosterveer and Sonnenfeld 2014). As Jouzi et al. (2017) report, the global food market for organic products has witnessed a remarkable growth from 15.2 billion USD in 1999 to72 billion USD in 2013. Although in several industrialised countries of the South, the rise of supermarket chains and a growing consumer demand is leading to a gradual expansion of the domestic market for organic products, the majority of organic food sales continue to take place in the United States and countries of the European

Union (Siricix et al. 2011). With demand for organic food rapidly increasing over the last two decades, organisations such as FAO and IFAD consider the organic industry as a robust market with huge development opportunities for smallholder farmers (Scialabba 2007).

The economic benefits that farmers gain from the organic market have been widely reported. Because organic agriculture relies on the use of local and natural resources, it can bring production costs down and increase financial security through productivity optimisation and yield stabilisation (Setboonsarng 2006). Studies demonstrated that farmers can make better savings by reducing the cost of input through substituting pesticides with locally grown natural inputs, as well as generate more employment opportunities given that organic production is more labour-intensive method (Maertens and Swinnen 2009; Rundgren and Parrott 2006; Van Elzakker and Eyhorn 2010). Moreover, access to foreign organic markets presupposes the participation of the Southern farmers in organic certification schemes that provides a guarantee for the pesticide-free cultivation of the crops and for the required standard of quality. Organic certification can, therefore, provide the ticket for markets with higher profits due to the price premiums of organic products. In fact, with price premiums for organic products ranging between 10 and 300 per cent, organic farmers can receive up to 50 per cent of this price premium, which is believed to increase the contribution of organic farming to economic development and poverty alleviation (Setboonsarng 2006). In a study on organic pineapple farming in Ghana, Kleemann et al. (2014) examined the benefits that farmers gained from organic certification in terms of return on investment. Their findings revealed that because the higher prices for organic products compensate for the lower yield, the overall profit from growing and selling organically certified fruits was higher than conventional farming, with a higher return on investment. In another comparative study conducted in India, Ramesh et al. (2010) showed that despite a decrease in crop productivity, farmers still managed to secure a higher profit from organic farming mainly thanks to the price premiums and the reduced cost of production. Organic farming can also provide opportunities for new partnerships, higher bargaining power, and better access to credits (Shah et al. 2003; Kilcher 2007; Bolwig et al. 2009).

The environmental and economic benefits that farmers reap through their participation in the organic movement are reflective of an alternative or counter-discourse against the exploitative structures of the global food market. Yet, this is only one side of the story, given the movement's co-optation into an uneven neoliberal global food market. Most studies that have looked into the challenges of organic agriculture for small-scale farmers have highlighted the inability of many farmers to produce sufficient quantities of organic compost, the long conversion periods, and the decreasing incomes during these periods that pose significant financial risks for producers. Farmers must farm organically for at least 3 years before they can be certified and before their products can be deemed organic. This is a challenging period that demands increased financial and time investment without the sufficient returns since farmers cannot sell at higher prices until they have been certified (Lotter 2015; Seufert 2012). A major constraint lies in the fact that organic agriculture in poorer countries of the South is driven primarily by consumer demand in the developed North, given the small internal demand. The high expectations and strict quality standards mean that only a small number of farmers can afford to gain access to foreign markets and export high-value agricultural products. The lack of a unifying single standard means that farmers need to meet multiple standards set by the governments of different countries. Although the EU and the USA both have single standards for organic products, countries like Japan, China, Australia, or Brazil enforce separate standards that producers have to adopt in order to access these markets, a process that involves higher inspection and auditing costs (IFOAM 2016).

Apart from the high certification costs that farmers are often unable to afford, only few countries in the South meet the necessary requirements and high certification expectations, have made arrangements to have their products recognised as organic by the US and European markets, and are therefore able to participate in these commodity networks (Kim 2013; Raynolds 2004). The different regulation systems for organic food across the world render the certification process a highly complicated and risky undertaking, with Southern producers having no negotiating power to request more realistic expectations or even have a dialogue with certifiers. Small producers are often unable to fulfil the quality criteria and standards required for export, which, coupled with the lack of the necessary infrastructure and the increasing costs of transport, increases the risk of them becoming excluded from the market (Jaffe 2007; Oosterveer and Sonnenfeld 2014).

Eventually, the organic trade risks re-enforcing the subordination of Southern producers who are excluded from decision-making processes regarding the legislation and standards for organic production. In this case, organic trade and the certification process more specifically exhibit symptoms of a top-down modernisation approach. Although organic agriculture encourages the use of traditional knowledge around indigenous methods, farmers still need to be educated on the certification process, and once again the knowledge transfer process remains linear and one-directional. Research and knowledge generation on organic farming is conducted predominantly in developed countries where the needs and knowledge of farmers is neglected (HLPE 2013; Seufert et al. 2012). More crucially, standards of production and processing of organic crops are developed in countries of the North with environmental and cultural conditions that differ dramatically to the regions in the South and with an entirely different producer in mind. When this model is transferred uncritically to regions with different local conditions, a new form of neo-colonialism emerges, reinforcing the colonial subservience of the Southern producer (Jaffee 2014; Raynolds 2004). Not only are Southern producers excluded from negotiations with buyers and decisions around prices and certification criteria but also assistance and development is offered to them even if it is not requested or without prior knowledge of what help is really needed (Dolan 2005; Mirowski 2001). These problems are exacerbated by the rigid environmental and production criteria and standards that are designed in the developed North and imposed on farmers in the South, lacking the local knowledge that is required to make them applicable to heterogeneous ecosystems (Scott 1988 in Moberg 2016). Such practices are in line with the diffusion of innovation paradigms that see the transfer of Western knowledge to producers in the South with the aim to achieve progress and modernisation.

The continuous exploitation of organic farmers by middlemen and corporations reproduces a form of colonial exporting relations between farmers, buyers, and consumers that harms the legitimacy of organic trade as a socially sustainable development model. The exploitation of marginalised farming communities by middlemen and corporate buyers is encouraged by the mainstreaming of the organic trade and the concentration of buying power in supermarkets and food conglomerates that set and control the quality standards in the global food market. In fact, in a global market where organic products represent only 2 per cent of food and drink sales, supermarkets have the dominant share of the organic food market in Europe, with certain supermarket chains, such as the UK-based Sainsbury's introducing their own organic brands, while conglomerates including Mars, Heinz, Nestle, and Danone have all launched their own organic lines (Sylvander and Kristensen 2004). The corporate dominance of organic foods has led to new institutional arrangements for the certification of organic products, encouraging a standardisation of product requirements and reducing the organic label to a generic production claim (Goodman et al. 2014; Henson and Reardon 2005). Such overcrowding of the organic market has ultimately encouraged further competition and exploitation of producers, with retailers demanding lower prices and larger volumes from suppliers, essentially reproducing the standards of conventional agricultural production. Ironically, at the same time, and despite standardisation of quality specifications, corporate retailers remained picky and demanding, posing additional challenges to smaller and mixed-crop food growers (Smith 2006).

Effectively, although the organic movement is promoted as a corrective to the growing environmental and trade concerns, the opportunities for resistance to the damaging inequalities of the global food economy can be significantly curtailed by the very same neoliberal relations that have come to define the global organic food market. But a key observation to be drawn from the review of the studies presented in this section is that ethical and environmental practices can both co-exist and be co-opted by the global market. In this way, the organic movement is reflective of the complex structure of the global food economy and the multidimensional and dynamic nature of food production and consumption practices that are underpinned both by traditional market forces and ethical concerns. This complexity becomes even more pronounced in the case of fair trade that has sparked debates about the (in)compatibility of development and the market. However, as with the organic movement, fair trade has planted the seed for a different understanding of sustainable development that can emerge from the structures and relations that make up the global food economy.

Fair Trade and Development

The fair trade movement is considered as the archetypal globalised Alternative Food Network (AFN), with a history that spans over six decades, since it first arose after WWII. As Goodman et al. (2014) recount, the history of fair trade has already been told through a rich literature that examines the journey of the movement from its origins and inception until today (see the work of Anderson 2009a, b; Bacon 2010; Jaffee 2014: Fridell 2007; Kleine 2005; Levi and Linton 2003; Lyon and Moberg 2010; Hudson et al. 2013; Nichols and Opal 2005; Raynolds et al. 2007;

Renard 1999, 2003, 2005; Schmelzer 2010). Rather than offering another detailed historical account, this chapter positions the fair trade movement in the historical context of the neoliberal global food market that has shaped the movement's development ideology. Fair trade has been caught in a paradox and a tension between its economic and solidarity-driven approach to development that has raised questions about the extent to which it truly fulfils its promise of the protection of the farmers from the forces of the global food economy. But evaluating fair trade strictly as an alternative to commodity capitalism and mainstream markets has now reached an impasse, leading to new questions among AFNs scholars about the purpose and value of the movement for farming communities (Naylor 2018).

The fair trade movement has its origins in the same period as the emergence of neoliberalism and the creation of the Bretton Woods institutions, that is, the World Bank, the IMF, and the General Agreement on Tariffs and Trade (GATT), and embodied a resistance to the deregulation promoted by neoliberal policies (Fridell 2007; Jaffee 2014). This was a time during which the United Nations promoted their agenda for more equitable trade relations between the countries in the Global North and the Global South in an effort to address the South's reliance on primary product exports and secure fairer prices for primary product exporters of the South (Lyon and Moberg 2010).

From its origin, the fair trade movement has encompassed at least two lines of focus: the development and the solidarity line. The focus on development was initially embraced by religious and secular solidarity organisations and charities both in the United States and Europe[1]—also known as Alternative Trading Organizations or ATOs—that sought to open up export markets for the products, mainly handicrafts, of impoverished and disadvantaged people. As this alternative trade continued to grow, a movement began to take shape. The development remit was clearly linked to the income generation and economic growth that was achieved through sales, but in its initial stages, fair trade also incarnated a political tradition that found expression in the solidarity of many new trading groups—for example, Twin Trading in the UK, Oxfam-Wereldwinkels in Belgium, and Stichting Ideele Import in the Netherlands—towards socialist countries

[1] These included Ten Thousand Villages, Equal Exchange (United States), Oxfam, Christian Aid (UK), GEPA (Germany), and Max Havelaarin (Netherlands) (Goodman et al. 2014: 197).

that were disadvantaged by US import policies and product embargos. It was through these solidarity groups and networks that fair trade began to be linked to a larger critique towards neoliberal policies and the capitalist system (Jaffee 2014: 38–40).

The fair trade system is predicated on the creation of more socially and environmentally responsible trade relations between food growers in the South and consumers in the North primarily through a number of measures such as the removal of intermediaries, the sourcing of products from democratically organised cooperatives, the higher prices paid by consumers, and the social premium paid on top of the selling price that farming communities can invest in development projects. A labelling and certification system ensures that the products sold through fair trade channels are produced and sold under the equitable relations and fair conditions stipulated by the different fair trade organisations.

The social premium generated by fair trade is a central component of the system, as it contributes to a much-desired fair wage for producers. At the same time, it is cast as an alternative funding mechanism to government support, especially given the struggles that many indebted governments in the Global South are facing (Moberg 2016). In fact, neoliberal institutions including the World Bank consider fair trade as preferable to Keynesian models of state-driven solutions, and promote it as an alternative to the national government policies that fail to protect their farmers (Fridell 2007; Harvey 2007; Lyon and Moberg 2010). Admittedly, only in 2016, fair trade producers worldwide received more than €150 million generated through social premiums (Fairtrade Foundation), a figure that reflects the rapid expansion of fair trade and the volume of funding that comes with it. Social premiums embody values of social justice and environmental protection that liken fair trade to a 'Polanyian countermovement' and a form of civil society that is expected to 'step in' when government and the state are unable to meet social needs (Guthman 2007; Raynolds 2000; Wilson and Mutersbaugh 2015). The market-access approach to fair trade's development cannot really be separated from the market-reform stance that supports efforts '…to penetrate the market and the lifestyle of consumers in order to sell larger quantities of fair trade products' (Renard 2003: 91).

The role of the consumer is another manifestation of fair trade's cooptation of the market logic. Consumption, as a form of collectivised politics, lies at the heart of fair trade. Ordinary people are enlisted into social change projects through their purchase of ethical goods while sales are translated into development policies (Barnett et al. 2005: 39; Goodman et al. 2014). Here, fair trade marketing and advertising deploy the notion of the sovereign consumers that, through the construction and transformation of their social and political identities, have the power to transform the market and society too (Lyon 2006). Appadurai refers to the fetishism of the consumers who believe they are agents (Appadurai 1990: 307), a belief that disguises the consolidation of markets in the hands of fewer corporations (Lyon and Moberg 2010). This form of 'developmental consumption' (Goodman 2010) has become fair trade's primary pathway to environmental stewardship and social justice through an alleged producer-consumer partnership based on equal mutual benefit at both ends of the chain (Ashman 2001). Such equity is of course illusionary. On the one hand, through the certification seal and the shortening of the commodity chain, fair trade embraces a closer connection between food growers and consumers, and the use-value of fair trade products is supposed to derive from the utility of the products for the consumer and the fair price for the producer. Yet, as products still must pass through importers and/or roasters, with each taking a cut, the empowerment of the farmers through the choices of benevolent consumers is much less than imagined (Jaffee 2014). On the other hand, while consumers' 'labour' comes in the form of choice, for producers, the investment and participation in the burdensome and expensive certification systems is a matter of survival, since this is the only way for them to access export markets (Goodman et al. 2014: Melo and Hollander 2013). Eventually, farmers' economic survival becomes dependent on volatile luxury food commodity markets where the investment costs often outweigh the benefits. In a study of Ecuadorian cocoa producers, Melo and Hollander (2013) demonstrated the often-excessive risks to which market-based development programmes expose farming communities in the rural South. In the case of the Ecuadorian producers, the drastic fluctuations of cocoa prices made them vulnerable to the volatility of the global food market, while their specialised knowledge and the uniqueness and heritage of their products is essentially rendered invisible in the workings of corporations such as Kraft. Moberg's (2016) study of fair trade in Dominica exposed the failure of social premium to protect Caribbean banana producers against the competitive market environment dominated by corporations of the likes of Chiquita.

Finally, like organic trade, the 'entrepreneurial developmentalism' of fair trade is compromised by its mainstreaming and marketisation. In practice, this marketisation is manifested in the placing of fair trade products on the shelves of supermarkets alongside mainstream brands, and the inclusion of food multinationals in the movement. One of the key moments in the movement's mainstreaming was the entrance of Starbucks into the sourcing and selling of fair trade coffee, a move that paved the way for the inclusion of more mainstream multinationals into the game. Another key moment of this commercialisation is encapsulated in the entrance of two of UK's leading fair trade brands, Cafe' Direct and Divine Chocolate, into the mainstreaming market. This was a key moment also for the success and growth of the movement and for changing the attitudes of consumers (Golding and Peattie 2005; Low and Davenport 2005; Moore et al. 2006; Nichols and Opal 2005). The neoliberal economic model and corporate consolidation that drove this move was coupled with the emergence of a private regulatory system that saw the establishment of Fairtrade Labelling Organizations (FLO) in 1997, and with this, the arrival of new product certification and labelling rules that paved the way for more mainstream commercial enterprises to enter fair trade (Dolan 2010; Low and Davenport 2005).

Although the engagement of fair trade with the mainstream market also marked its deviation from its charity status and logic of alternative exchange, it was seen by both activists and brands as a unique opportunity to share the benefits of fair trade with more producers through a growing market share and sales. More importantly, this was an attempt to re-introduce fair trade as 'entrepreneurial developmentalism' (Dolan 2010) that shifted the emphasis from charity and aid dependency of food growers and producers to a more equitable market-driven exchange. For this model to work, a key strategy was the so-called 'quality' turn that saw fair trade products being repositioned in the market as 'better quality' and 'better taste'. This was a strategy that put the mainstream consumer in the driving seat, as it was through the growth of sales and market expansion that the benefits of fair trade would be spread even further. In a way, the quality turn also cemented fair trade as an expression of development through consumerism and consumer choice. Mainstreaming may have brought more development in the form of a growing market share, but it has aggravated the risk of fair trade becoming part of the same corporate logic that it was meant to oppose (Goodman et al. 2014).

Needless to say that the logic of development through market expansion has been widely contested by scholars who point to the unfair competitive advantage that corporations and mainstream brands enjoy, thanks to the work done by ATOs to establish certification rules and build the fair trade brand. Part of this advantage is the use of the 'halo effect' of ethical branding developed by ATOs and the introduction of tougher new regulations for quality requirements and certification (Dolan 2008; Low and Davenport 2007; Reed 2009). Others are also critical of the consequences of fair trade's mainstreaming for small producers. For instance, Tallontire (2000) studied the impact that the liberalisation of coffee exports had for a cooperative of coffee growers in the Kilimanjaro region of Tanzania, KNCU, especially following their partnership with Cafe' Direct and its founding company, Twin. The case of KNCU summarises some of the threats and challenges that farmers and cooperatives must face. On the one hand, KNCU had a regular buyer and were able to increase exports of its members' produce; on the other, it was also in direct competition with many other buyers for this coffee. Moreover, KNCU members had minimal to no contribution to the strategy of the trading networks nor were they assisted in developing their own marketing strategy. Many of the challenges were also the result of internal tensions and slow change in the cooperative; but the case also demonstrates the one-dimensional benefits that a sales and consumerism-oriented approach to development can provide.

Critics of the fair trade movement point to a deep paradox and dilemma. It embraces social justice and environmental sustainability and seeks to encourage an opposition to the effects of neoliberal globalisation and alter the unfair trade relations that disadvantage small farmers in the Global South. However, the solution it offers is predicated on the premises of the very markets that have created the problem, since in order for its benefits to grow, the fair trade system has retained several of the mainstream players and practices found in the conventional commodity chain, such as large corporations and large-scale plantations (Jaffee 2014; Lyon and Moberg 2010). Although fair trade is supposed to serve a different role vis-à-vis capitalism and the market, its development and social justice remit is inseparable from capitalism and its role in facilitating access for Southern producers to the Northern markets. Market access is seen as a pathway to development, poverty eradication, and justice, as well as to building producer capacity for more successful integration in the global food economy (Fridell 2007; Jaffee 2014).

Yet, there is another caveat: framing fair trade as a platform of alternative economic exchanges and capitalism's potential other ends up reinforcing capitalism's discursive hegemony. This conceals the complex, messy, and dynamic nature of the economic and social practices that constitute the fair trade movement and AFNs more broadly (Healy 2009; Naylor 2018). These practices are based on independencies among multiple actors and require new analytical approaches and new ways of producing knowledge about economic exchanges. In search of new analytical approaches, AFNs scholars have begun to turn to alternative economic theories to uncover the multitude of relations and associations involved in agriculture and food production. These approaches embody a new, alternative language and new line of thought around development in global food networks. This is the type of language that can pave the way for an alternative understanding of sustainable development.

Towards a New Language of Development in the Global Food Economy

In search for new approaches to the economic practices that constitute AFNs, scholars have turned to the work of economic geographers Katherine Gibson and Julie Graham on diverse economies, calling for more attention to the heterogeneous economic activities and multiple 'ecologies of productivity that permeate alternative food networks' (Harris 2009; Maye and Kirwan 2010; Naylor 2018). Naylor (2018) uses the framework of diverse economies to demonstrate the diversity of economic exchanges and social formations that underpin community economic spaces, and the multiplicity of connections and relations that exist across the fair trade commodity chain, and which span beyond the producer-consumer relation. Studies like Naylor's draw attention to the need for deeper consideration of the multiplicity of actors, nodes, and relations in the food commodity chain, and the different development possibilities these can generate for communities in the Global South. What tends to be forgotten is that sustainable development and social change is a highly nuanced process that is contingent on people's everyday realities rather than just a fight against an abstract global industrial food system (Arce et al. 2015). An understanding of 'resistance' should be searched for in the multitude of relations and associations involved in agriculture and food production (Narvaez Mena 2017). Through this framing, food production is more than the sum of resources, and is born from an array of human and non-human relations and social networks (Carolan 2015; Narvaez Mena 2017). At the same time, this relational approach requires a reconceptualisation of the traditional economic hierarchy between states, markets, and farmers, not just as one of dominance but as a co-creative one (Ofstehage 2017).

In the case of global food networks, the network of organisations working under the principles of fair trade, for instance, are bound together by a common social and environmental goal and cause, which sets a common understanding of the rules and norms of their collaboration. This allows even loosely connected organisations and companies to share information based on trust and cooperate on areas such as marketing and sales. The case of Equal Exchange and Divine Chocolate is a useful example of two fair trade companies that share information on a regular basis, thanks to close relations and friendship that bind their members (Davies and Ryals 2010). However, these non-monetary relations have received less attention in the context of development and the global economy despite the potentially valuable input that communities can reap from them in their market exchanges. For instance, in the case of the Wayanadan farmers' cooperative, the farmers' network of partners consists of a group of private SMEs from Europe that are bound by close, trusted relations, and in some cases even by friendships, which contribute significantly to the farmers' development. These relations have led not just to the sharing of information among the network but also to joint organisation of promotional events for the farmers' products that have brought economic benefits for all parties and have contributed to the farmers' well-being more broadly (see Chap. 4).

The explanatory power of Gibson-Graham's work, in terms of theorising the relational aspect of the global food economy, lies in the new language and new way of thinking about the economy that it provides. By doing so, the language of Gibson-Graham's diverse economies offers a tool for understanding the rich diversity of economic practice and development opportunities in the global food economy and in AFNs. In *A Postcapitalist Politics*, one of their most influential works, they describe the significance of language as follows:

> A language is fluid and mobile, not easily confined to a particular location or scale ... it provides the contours and emphases on other worlds, but cannot tell you what to say. It can share the space of power with other languages without having to "overthrow" them (Gibson-Graham 2006: xxxiv).

Gibson-Graham's pursuit of a new language of the economy was driven by Laclau and Mouffe's theory that discourse and language is at the centre of any political project, an idea that is also reflected in Chambers' proposition to bring forward and create space for the language of alternative disciplines (Chambers 1997). Gibson-Graham advance the idea that language contributes to the performativity of knowledge and that we actively participate in constituting our object of study through the choice of language. Their theory and concept of diverse economies has been inspired by a feminist post-structural approach that a language that promotes a totalising conceptualisation of capitalism inevitably conceals the diversity of interests and motivations that drive economic activity, other than profit maximisation and individualistic economic rationality. As a result, talking about the economy in certain ways and using hegemonic discourses encourages the reproduction of certain conceptions of capitalism and neo-liberalism as a uniform and totalising process. Inspired by feminism, post-structuralism and queer theory Gibson-Graham's work points out that global capitalism is not all-encompassing or the benchmark for any economic activity, but the outcome of many types of economic activities that are often less visible to researchers. Gibson-Graham (2014) are critical towards mainstream economic discourses that emphasise certain practices and motivations such as pursuit of profit, competition, efficiency, innovative entrepreneurship, and self-interest. For them, the economy is like an iceberg, where the tip constitutes formal market transactions, wage labour, and capitalist enterprises. These are underpinned by submerged non-market transactions and social relations, as well as altruism and solidarity. Although the activities and relations below the waterline are generally hidden from view, the sustainability of the economy depends on these to function. The transactions that constitute the economy are connected to 'other-than-capitalist' relations that sustain the iceberg (Gibson-Graham 2006: 71). Yet, as these are rarely represented or talked about, the dominance of capitalism as all-encompassing is strengthened. The authors highlight the much wider range of social relations that underpin economic practices, including trust, care, sharing, spiritual connection, and environmental and social justice, to name but a few.

At the core of Gibson-Graham's theory and new language of diverse economies is Eve Sedgwick's weak theory, a theory underpinned by the distinction between what she calls 'strong/universal' and 'weak/localised' theorising as a way to construct and perform knowledge. While knowledge in the social sciences is predominantly generated through strong theories, it is through weak and localised approaches that the world of social science can be better understood. Amongst the most inspirational ideas in Eve Sedgwick's work is her association of dominant influential theory with 'paranoia', a term that she borrowed from psychoanalyst Melanie Klein. In her theorising of affect, Sedgwick (2003: 128-130) explained how the way critics and scholars approached their objects of study was driven by continuous suspicion and fear of surprise. In her seminal work *Touching Feeling*, she paints an interesting image of the critical thinker, researcher, and investigator who strives to develop her theoretical and discursive framework in order to expand its reach and reductiveness. For Sedgwick, the paranoia with wide generality is what leads to the emergence of 'strong' theory. Inspired by Silvan Tomkins' more psychoanalytical approach, Sedgwick talks about the paranoia of social theorising to minimise surprise that yields a kind of strong theory where everything comes under a usually large concept such as neoliberalism, or globalisation. In Tomkins' words, strong theory always has something to say about everything; it accounts for a broad range of very remote phenomena, based on which its explanatory power is also evaluated. For Sedgwick, strong theory is another expression of paranoid reading that does not allow for anything new to be experienced. The main problem with this kind of thinking and knowledge production lies in reinforcing the status quo and the already 'known' while shutting down avenues to surprise and to alternative theoretical investigations.

The opposite of strong theory is 'weak' theory, which—according to Silvan Tomkins who coined the term—might be 'little better than a description of the phenomena which it purports to explain' (Sedgwick 2003: 134). Yet, this descriptive nature makes it no less powerful than strong theory. If there is one key difference, this is in its ability to offer richer, more focused, and localised descriptions and explanations. The merit of weak theory lies in exactly *not* trying to be inclusive and all-encompassing in its explanatory premises, and *not* trying to confirm what we already know. Instead, its purpose is to open new spaces of freedom and possibility for understanding and knowing the world. For Sedgwick, weak theory represents a 'reparative' model of reading that asks the reader, critic, and researcher to let go of the paranoid position and instead be open to experience surprise (Sedgwick 1997, 2003).

Following Sedgwick's thinking, the proposition of Gibson-Graham is that an economic language that 'signals the ever-present possibility of remaking economy in alternative terms' [Gibson-Graham 2006: xxxiv] is embodied in a weak theory that sees the economy through non-mainstream lenses and is populated by a variety of practices. Their theorisation of the 'weak economy' draws an explicit connection between Sedgwick's questions of how to rethink the economy and reconstitute capitalism as a multidimensional and diverse process. But other diverse economies scholars have also been inspired to understand their bodily dispositions—what Sedgwick describes as the paranoid 'anxiety-mitigating' stance—and thus to explore ways for generating different economic representations (Sedgwick 1997, 2003; Brown et al. 2011).

In this case, while strong theory would approach development and poverty eradication as reconcilable with top-down profit maximising solutions, a weak theoretical approach would be more attentive and adaptive to local conditions and help unearth the various development opportunities that reside in those hidden relations that underpin the global food economy. Moreover, Gibson-Graham's 'decomposition' of the economy enables an alternative conceptual and empirical connection between development, social change, and the global food economy creating space for understanding development and social change as an embodied experience contingent on people's everyday realities and relations. By urging us to follow queer theory's tools and read 'for difference rather than dominance', they are prompting us to bring into visibility the great heterogeneity that characterises capitalist practices and all of the non-capitalist activity that sustains the global food market. This includes an exploration of those less formal social relations that sustain economic transactions but remain hidden when one looks at the economy only through the prism of capitalism. In this way, apart from offering a new approach to the global food economy, Gibson-Graham's proposition provides a much needed theoretical tool that ties in with the need to bring into visibility the informal social relations and communication processes that contribute to sustainable development.

Conclusion

In today's neoliberal context characterised by a triple crisis—food, energy, and financial—concepts of sustainable development are in flux and issues of well-being are becoming more complex to explain (Van Hemelrijck 2013). The area of food and agriculture epitomises this complexity. Development and social change in agriculture and food is not the outcome of unidirectional development processes. Instead, it emerges from multiple socio-material relations and embodied subjectivities (Blackman et al. 2008; Woods 2007). These socio-material relations are also at the core of the global food economy, but they are only partially visible and studied, as the emphasis is disproportionately on economic relations and outcomes. The study and understanding of development in the global food economy is foregrounded by a discursive approach that is wedded to a dominant economistic ideology. Development is conceptualised through the integration of Southern farmers in the global market, while development efforts such as the Green Revolution reflect the dominant econocentric discourse that has been consistently promoted by governments and development institutions. Movements such as the organic and fair trade initiative are meant to offer alternative contexts and economic practices, but in reality they end up reinforcing the dominant discourses of development. Our knowledge is, therefore, governed by an economistic conception of development that is the resultant of the language and discursive approach that has dominated the field. This 'strong theory' approach proves questionable in its capacity to offer a comprehensive account and understanding of the nuances of sustainable development and social change in the context of the global food economy. Unpacking the development opportunities that reside in the global food economy requires a new language that will create space for a new interdisciplinary approach. In this book, the need for a new language is met by weak theory, and in the case of the food economy, the theory of diverse economies provides such a weak theory approach. By bringing into light the sociability of the economy, the theory of diverse economies creates space for the role of communication to be analysed as a component of the food economy and as a pathway to sustainable development. It provides a solid foundation for developing a new interdisciplinary approach to the global food economy and the development opportunities that exist in it, which is indispensable in deconstructing and re-politicising the idea of sustainable development.

References

Abrahams, C. (2007). Globally useful conceptions of alternative food networks in the developing south: The case of Johannesburg's urban food supply system. In D. Maye, L. Holloway, & M. Kneafsey (Eds.), *Alternative food geographies: Representation and practice* (pp. 95–114). Amsterdam: Elsevier.

Albersmeier, F., Schulze, H., Jahn, G., & Spiller, A. (2009). The reliability of third-party certification in the food chain: From checklists to risk-oriented auditing. *Food Control, 20*(10), 927–935. https://doi.org/10.1016/j.foodcont.2009.01.010.

Altieri, M. A. (2009). The ecological impacts of large-scale agrofuel monoculture production systems in the Americas. *Bulletin of Science, Technology and Society, 29*(3), 236–244. https://doi.org/10.1177/0270467609333728.

Anderson, M. (2009a). 'Cost of a cup of tea': Fair trade and the British co-operative movement, C. 1960–2000. In L. Black & N. Robertson (Eds.), *Consumerism and the co-operative movement in modern British history*. Manchester: Manchester University Press.

Anderson, M. (2009b). NGOs and fair trade: The social movement behind the label. In N. Crowson, M. Hilton, & J. McKay (Eds.), *NGOs in contemporary Britain: Non-state actors in society and politics since 1945*. London: Palgrave.

Appadurai, A. (1990). Disjuncture and difference in the global cultural economy. *Theory, Culture & Society, 7*(2), 295–310. https://doi.org/10.1177/026327690007002017.

Arce, A., Sherwood, S., & Paredes, M. (2015). Introduction: Food embodiments, assemblages and intersubjectivities: Ebbs and flows of critical food studies. In S. Sherwood, A. Arce, & M. Paredes (Eds.), *Food, agriculture and social change. The everyday vitality of Latin America* (pp. 1–20). New York: Routledge.

Ashman, D. (2001). Strengthening north-south partnerships for sustainable development. *Nonprofit and Voluntary Sector Quarterly, 30*(1), 74–98. https://doi.org/10.1177/0899764001301004.

Bacon, C. (2010). Who decides what is fair in fair trade? The Agri-environmental governance of standards, access and price. *Journal of Peasant Studies, 37*(1), 111–147.

Bakewell-Stone, P., Lieblein, G., & Francis, C. (2008). Potentials for organic agriculture to sustain livelihoods in Tanzania. *International Journal of Agricultural Sustainability, 6*(1), 22–36. https://doi.org/10.3763/ijas.2007.0266.

Barnett, C., Cloke, P., Clarke, N., & Malpass, A. (2005). Consuming ethics: Articulating the subjects and spaces of ethical consumption. *Antipode, 37*(1), 23–45.

Bazuin, S., Azadi, H., & Witlox, F. (2011). Application of GM crops in sub-Saharan Africa: Lessons learned from Green Revolution. *Biotechnology Advances, 29*(6), 908–912. https://doi.org/10.1016/j.biotechadv.2011.07.011.

Bernstein, H. (2000). *"The peasantry" in global capitalism: Who, where and why*. London: Merlin Press.

Biggs, S. D. (1990). A multiple source of innovation model of agricultural research and technology promotion. *World Development, 18*(11), 1481–1499. https://doi.org/10.1016/0305-750X(90)90038-Y.

Blackman, L., Cromby, J., Hook, D., Papadopoulos, D., & Walkerdine, V. (2008). Creating subjectivities. *Subjectivity, 22*(1), 1–27. https://doi.org/10.1057/sub.2008.8.

Bolwig, S., Gibbon, P., & Jones, S. (2009). The economics of smallholder organic contract farming in tropical Africa. *World Development, 37*(6), 1094–1104. https://doi.org/10.1016/j.worlddev.2008.09.012.

Brown, G., Browne, K., Brown, M., Roelvink, G., Carnegie, M., & Anderson, B. (2011). Sedgwick's geographies: Touching space. *Progress in Human Geography, 35*(1), 121–131.

Carolan, M. S. (2015). Adventurous food futures: Knowing about alternatives is not enough, we need to feel them. *Agriculture and Human Values, 33*(1), 141–152. https://doi.org/10.1007/s10460-015-9629-4.

Caron, P., Ferrero y de Loma-Osorio, G., Nabarro, D., et al. (2008). Food systems for sustainable development: Proposals for a profound four-part transformation. *Agronomy for Sustainable Development, 38*(4). https://doi.org/10.1007/s13593-018-0519-1.

Chambers, R. (1974). Managing rural development. *Institute of Development Studies Bulletin, 6*(1), 4–12. https://doi.org/10.1111/j.1759-5436.1974.mp6001002.x.

Chambers, R. (1981). Rapid rural appraisal: Rationale and repertoire. *Public Administration and Development, 1*(2), 95–106. https://doi.org/10.1002/pad.4230010202.

Chambers, R. (1997). Responsible well-being. A personal agenda for development. *World Development, 25*(11), 1743–1754.

Chambers, R., & Jiggins, J. (1987). Agricultural research for resource-poor farmers part I: Transfer-of-technology and farming systems research. *Agricultural Administration and Extension, 27*(1), 35–52.

Davies, I. A., & Ryals, L. J. (2010). The role of social capital in the Success of fair trade. *Journal of Business Ethics, 96*(2), 317–338. https://doi.org/10.1007/s10551-010-0468-3.

Dolan, C. (2005). Field of obligation: Rooting ethical sourcing in Kenyan horticulture. *Journal of Consumer Culture, 5*, 365–389.

Dolan, C. (2008). The mists of development: Fair trade in Kenya tea fields. *Globalizations, 5*(2), 1–14.

Dolan, C. (2010). Virtual moralities: The mainstreaming of fair trade in Kenyan tea fields. *Geoforum, 41*, 33–43.

Dutta, M. J. (2011). *Communicating social change: Structure, culture, agency.* New York: Routledge.

Egelyng, H., & Høgh-Jensen, H. (2006). Towards a global research programme for organic food and farming. In N. Halberg, H. F. Alroe, M. T. Knudsen, & E. S. Kristensen (Eds.), *Global development of organic agriculture: Challenges and prospects* (pp. 323–343). Oxford: CABI Publishing.

Escobar, A. (2001). *Encountering development: The making and unmaking of the third world.* Princeton: Princeton University Press.

Fanon, F. (1982). *The wretched of the earth.* New York: Grove.

Food and Agriculture Organization of the United Nations. (1996). *Rome declaration on world food security and world food summit plan of action.* Rome: World Food Summit.

Forssell, S., & Lankoski, L. (2015). The sustainability promise of alternative food networks: An examination through "alternative" characteristics. *Agriculture and Human Values, 32*(1), 63–75.

Freidberg, S., & Goldstein, L. (2011). Alternative food in the global south: Reflections on a direct marketing initiative in Kenya. *Journal of Rural Studies, 27*(1), 24–34. https://doi.org/10.1016/j.jrurstud.2010.07.003.

Fridell, G. (2007). *Fair trade coffee: The prospects and pitfalls of market-driven social justice.* Toronto: University of Toronto Press.

George, S. (1990). *Ill fares the land: Essays on food, hunger and power.* London: Penguin.

Gibson-Graham, J. K. (2006). *A postcapitalist politics.* Minneapolis: University of Minnesota Press.

Gibson-Graham, J. K. (2014). Rethinking the economy with thick description and weak theory. *Current Anthropology, 55*(9), 147–153.

Golding, K., & Peattie, K. (2005). In search of a golden blend: Perspectives on the marketing of fair trade coffee. *Sustainable Development, 13*(3), 154–165. https://doi.org/10.1002/sd.274.

Goodman, D. (2004). Rural Europe redux? Reflections on alternative agro-food networks and paradigm change. *Sociologia Ruralis, 44*(1), 3–16.

Goodman, M. K. (2010). The mirror of consumption: Celebritization, developmental consumption and the shifting cultural politics of fair trade. *Geoforum, 41*(1), 104–116. ISSN 0016-7185. https://doi.org/10.1016/j.geoforum.2009.08.003.

Goodman, D., DuPuis, M. E., & Goodman, M. K. (2014). *Alternative food networks: Knowledge, practice and politics.* New York: Routledge.

Gordillo, G., & Gómez, H. (2005). *Conversaciones sobre el hambre. Brasil y el derecho a la alimentación.* México, Centro de Estudios para el Desarrollo Rural y Sustentable y la Soberanía Alimentaria (CEDRSSA), Cámara de Diputados, LIX Legislatura/Congreso de la Unión.

Gordillo, G., & Jerónimo, O. M. (2013). *Food security and sovereignty.* Rome: Food and Agriculture Organization of the United Nations (FAO).

Guthman, J. (2007). The Polyanian way? Voluntary food labels as neoliberal governance. *Antipode, 39*(3), 457–478.

Gutiérrez-Escobar, L. (2018). Food sovereignty. In A. Kothari, A. Salleh, A. Escobar, F. Demaria, & A. Acosta (Eds.), *Pluriverse: A post-development dictionary* (pp. 185–188). New Delhi: Tulika Books.

Harris, E. (2009). Neoliberal subjectivities or a politics of the possible? Reading for difference in alternative food networks. *Area, 41*(1), 55–63.

Harvey, D. (2007). *A brief history of neoliberalism.* Oxford: Oxford University Press.

Healy, S. (2009). Alternative economies. In N. Thrift & R. Kitchin (Eds.), *International Encyclopedia of human geography: A 12-volume set* (pp. 338–344). Amsterdam: Elsevier Science.

Henderson, R., & Clark, K. (1990). Architectural innovation: The reconfiguration of existing product technologies and the failure of established firms. *Administrative Science Quarterly, 35*(1), 9–30. https://doi.org/10.2307/2393549.

Henson, S., & Reardon, T. (2005). Private Agri-food standards: Implications for food policy and the Agri-food system. *Food Policy, 30*, 241–253.

Hewitt de Alcántara, C. (1976). Modernizing Mexican agriculture: socioeconomic implications of technological change 1940–1970, Nr. 11; United Nations Research Institute for Social Development. Report, 76,5.

HLPE. (2011). *Price volatility and food security.* A report by the High Level Panel of Experts on Food Security and Nutrition of the Committee on World Food Security, Rome.

HLPE. (2013). *Investing in smallholder agriculture for food security.* A report by the High Level Panel of Experts on Food Security and Nutrition of the Committee on World Food Security, Rome.

House of Commons (2009) Securing food supplies up to 2050: the challenges faced by the UK - Environment, Food and Rural Affairs Committee, Fourth Report of Session 2008–09. Available at: https://publications.parliament.uk/pa/cm200809/cmselect/cmenvfru/213/213i.pdf

Hudson, I., Hudson, M., & Fridell, M. (2013). *Fair trade, sustainability and social change.* Basingstoke: Palgrave Macmillan.

IFOAM. (2016). *Power by the people.* Consolidated Annual Report of IFOAM, Organics International, available.

Ilbery, B., & Kneafsey, M. (1998). Product and place: Promoting quality products and services in the lagging rural regions of the European Union. *European Urban and Regional Studies, 5*(4), 329–341.

Ilbery, B., & Maye, D. (2005). Alternative (short) food supply chains and specialist livestock products in the Scottish–English borders. *Environment and Planning A, 37*, 823–844.

Ishii-Eiteman, M. (2009). Food sovereignty and the international assessment of agricultural knowledge, science and Technology for Development. *Journal of Peasant Studies, 36*(3), 689–700.

Jaffee, D. (2007). *Brewing justice: Fair trade coffee, sustainability, and survival.* Berkeley, CA: University of California Press.

Jaffee, D. (2014). *Brewing justice: Fair trade coffee, sustainability and survival.* Berkeley: University of California Press.

Jouzi, Z., Azadi, H., Taheri, F., Zarafshani, K., Gebrehiwot, K., Van Passel, S., & Lebailly, P. (2017). Organic farming and small-scale farmers: Main opportunities and challenges. *Ecological Economics, 132*(C), 144–154. https://EconPapers.repec.org/RePEc:eee:ecolec:v:132:y:2017:i:c:p:144-154.

Kilcher, L. (2007). How organic agriculture contributes to sustainable development. *Journal of Agriculture and Rural Development in the Tropics and Subtropics, 89*, 31–49.

Kim, M. (2013) Growing Organic around the World: Domestic Regulations, International Relations and Developing Countries' Involvement in Organic Production, paper presented at the Annual Meeting of the International Studies Association (ISA), San Francisco, CA.

Kleemann, L., Abdulai, A., & Buss, M. (2014). Certification and access to export markets: Adoption and return on investment of organic-certified pineapple farming in Ghana. *World Development, 64*, 79–92. https://doi.org/10.1016/j.worlddev.2014.05.005.

Kleine, D. (2005). fairtrade.com versus fairtrade.org: How fair trade organisations use the internet. *Interacting with Computers, 17*, 57–83.

Kneafsey, M., Cox, R., Holloway, L., Dowler, E., Venn, L., & Tuomainen, H. (2008). *Reconnecting consumers, producers and food: Exploring alternatives.* Oxford: Berg.

Lamine, C. (2005). Settling shared uncertainties: Local partnerships between producers and consumers. *Sociologia Ruralis, 45*(4), 324–345.

Leeuwis, C., & Aarts, N. (2011). Rethinking communication in innovation processes: Creating space for change in complex systems. *The Journal of Agricultural Education and Extension, 17*(1), 21–36. https://doi.org/10.1080/1389224X.2011.536344.

Levi, M., & Linton, A. (2003). Fair trade: A cup at a time? *Politics and Society, 31*(3), 407–432. https://doi.org/10.1177/0032329203254862.

Lipton, M. (2017). Staples production. Efficient "subsistence" smallholders are key to poverty reduction, development, and trade. In P. Pingali & G. Feder (Eds.), *Agriculture and rural development in a globalizing world: Challenges and opportunities* (pp. 82–103). London/New York: Routledge.

Lotter, D. (2015). Facing food insecurity in Africa: Why, after 30 years of work in organic agriculture, I am promoting the use of synthetic fertilizers and herbicides in small-scale staple crop production. *Agriculture and Human Values, 32*(1), 111–118. https://doi.org/10.1007/s10460-014-9547-x.

Low, W., & Davenport, E. (2005). Postcards from the edge: Maintaining the 'alternative' character of fair trade. *Sustainable Development, 13*, 143–153.

Low, W., & Davenport, E. (2007). Mainstreaming fair trade: Adoption, assimilation, appropriation. *Journal of Strategic Marketing, 14*(4), 315–327.

Lyon, S. (2006). Evaluating fair trade consumption: Politics, defetishization and producer participation. *International Journal of Consumer Studies, 30*(5), 452–464.

Lyon, S., & Moberg, M. (Eds.). (2010). *Fair trade and social justice: Global ethnographies.* New York: New York University Press.

Maertens, M., & Swinnen, J. (2009). Trade, standards, and poverty: Evidence from Senegal. *World Development, 37*(1), 161–178.

Marsden, T., & Smith, E. (2005). Ecological entrepreneurship: Sustainable development in local communities through quality food production. *Geoforum, 36*(4), 440–451.

Marsden, T., Flynn, A., & Harrison, M. (2000). *Consuming interests: The social provision of food.* London: UCL Press.

Marsden, T., Banks, J., & Bristow, G. (2002). The social management of rural nature: Understanding agrarian-based rural development. *Environment and Planning A, 34*, 809–825.

Maye, D., & Kirwan, J. (2010). Alternative food networks. *Sociopedia.isa.* https://doi.org/10.1177/205684601051.

Maye, D., & Kirwan, J. (2013). Food security: A fractured consensus. *Editorial for the Journal of Rural Studies Special Issue: 'Food Security: Emerging Perspectives and Responses', 29*(1), 1–6.

McEwan, C. (2008). *Postcolonialism and development.* London: Routledge.

McMichael, P. (2005). Global development and the corporate food regime. *Research in Rural Sociology and Development, 11*, 265–299. https://doi.org/10.1016/S1057-1922(05)11010-5.

McMichael, P. (2011). Food system sustainability: Questions of environmental governance in the new world (dis)order. *Global Environmental Change, 21*(3), 804–812.

Melkote, S. R., & Steeves, H. L. (2015). *Communication for development: Theory and practice for empowerment.* New Delhi: SAGE.

Melo, C. J., & Hollander, G. M. (2013). Unsustainable development: Alternative food networks and the Ecuadorian Federation of Cocoa Producers, 1995–2010. *Journal of Rural Studies, 32*(1), 251–263. https://doi.org/10.1016/j.jrurstud.2013.07.004.

Millner, N. (2016). Food sovereignty, permaculture and the post-colonial politics of knowledge in El Salvador. In M. Wilson (Ed.), *Postcolonialism, Indigeneity and struggles for food sovereignty: Alternative food networks in the postcolonial world.* London: Routledge.

Mirowski, P. (2001). Refusing the gift. In S. Cullenberg, J. Amariglio, & D. Ruccio (Eds.), *Postmodernism, economics and knowledge* (pp. 431–458). London: Routledge.

Moberg, M. (2016). Market's end: Fair-trade social premiums as development in Dominica. *American Ethnologist, 43*(4): 677–690.

Moore, G., Gibbon, J., & Slack, R. (2006). The mainstreaming of fair trade: A macro-marketing perspective. *Journal of Strategic Marketing, 14*, 329–352.

Murray McIntire, J. (2017). Origins and some empirical tests of theories of agricultural evolution. In P. Pingali & G. Feder (Eds.), *Agriculture and rural devel-*

opment in a globalizing world: Challenges and opportunities (pp. 130–161). London/New York: Routledge.

Narvaez Mena, E. H. (2017). Affective struggles in the desert: Bringing water to bear on agriculture and food. In S. Sherwood, A. Arce, & M. Paredes (Eds.), *Food, agriculture, and social change: The everyday vitality of Latin America* (pp. 73–86). London/New York: Routledge.

Naylor, L. (2018). Fair trade coffee exchanges and community economies. *Environment and Planning A: Economy and Space, 50*(5), 1027–1046. https://doi.org/10.1177/0308518X18768287.

Nemes, N. (2009). *Comparative analysis of organic and non organic farming systems: A critical assessment of farm profitability.* Rome: FAO.

Nichols, A., & Opal, C. (2005). *Fair trade: Market-driven ethical consumption.* London: Sage.

Ofstehage, A. (2017). Encounters with the Brazilian soybean boom: Transnational farmers and the Cerrado. In P. Pingali & G. Feder (Eds.), *Agriculture and rural development in a globalizing world: Challenges and opportunities* (pp. 60–72). London/New York: Routledge.

Oosterveer, P., & Sonnenfeld, D. (2014). *Food, globalization and sustainability.* London/New York: Routledge.

Pachón-Ariza, F. A. (2013). Food sovereignty and rural development: Beyond food security. *Agronomía Colombiana, 31*(3), 362–377.

Parrott, N., & Marsden, T. (Eds.). (2002). *The real Green Revolution: Organic and agroecological farming in the south.* London: Greenpeace Environment Trust.

Patel, R. (2009). What does food sovereignty look like? *Journal of Peasant Studies, 36*(3), 663–706.

Perkins, J. H. (1997). *Geopolitics and the Green Revolution: Wheat, genes, and the cold war.* New York: Oxford University Press.

Pingali, P. L. (2012). Green Revolution: Impacts, limits, and the path ahead. *Proceedings of the National Academy of Sciences, 109*(31), 12302–12308. https://doi.org/10.1073/pnas.0912953109.

Pingali, P. L., & Feder, G. (2017). Introduction. In P. Pingali & G. Feder (Eds.), *Agriculture and rural development in a globalizing world: Challenges and opportunities* (pp. 1–8). London/New York: Routledge.

Ramesh, P., Panwar, N. R., Singh, A. B., Ramana, S., Yadav, S. K., Shrivastava, R., & Subba Rao, A. (2010). Current science, status of organic farming in India. *Current Science, 98*(9), 1190–1194.

Raynolds, L. T. (2000). Re-embedding global agriculture: The international organic and fair trade movements. *Agriculture and Human Values, 17*, 297–309.

Raynolds, L. T. (2004). The globalization of organic agro-food networks. *World Development, 32*, 725–743.

Raynolds, L., Murray, D., & Wilkinson, J. (Eds.). (2007). *Fair trade: The challenges of transforming globalization.* London: Routledge.

Reed, D. (2009). What do corporations have to do with fair trade? Positive and normative analysis from a value chain perspective. *Journal of Business Ethics, 86*, 3–26.

Renard, M. C. (1999). The interstices of globalization: The example of fair trade coffee. *Sociologia Ruralis, 39*(4), 484–500.

Renard, M. C. (2003). Fair trade: Quality, market and conventions. *Journal of Rural Studies, 19*(1), 87–96.

Renard, M. C. (2005). Quality certification, regulation and power in fair trade. *Journal of Rural Studies, 21*, 419–431.

Renting, H., Marsden, T., & Banks, J. (2003). Understanding alternative food networks: Exploring the role of short food supply chains in rural development. *Environment and Planning A, 35*, 393–411.

Rundgren, G. and Parrott, N., (2006). *Organic agriculture and food security*. IFOAM.

Schanbacher, W. (2010). *The politics of food: The global conflict between food security and food sovereignty*. Santa Barbara: Praeger.

Schmelzer, M. (2010). Marketing morals, moralizing markets: Assessing the effectiveness of fair trade as a form of boycott. *Management and Organizational History, 5*(2), 221–250.

Scialabba, N. (2007, May 3–5). Organic agriculture and food security. In *International Conference on Organic Agriculture and Food Security*, Food and Agriculture Organization of the United Nations, Italy. (OFS/2007/5).

Scott, J. W. (1988). *Gender and the politics of history*. New York: Columbia University Press.

Sedgwick, E. K. (1997). Paranoid reading and reparative reading; or, You're so paranoid, you probably think this introduction is about you. In E. K. Sedgwick (Ed.), *Novel gazing: Queer readings in fiction* (pp. 1–37). Durham: Duke University Press.

Sedgwick, E. K. (2003). *Touching feeling: Affect, pedagogy, performativity*. Durham/London: Duke University Press.

Setboonsarng, S. (2006). *Organic agriculture, poverty reduction, and the millennium development goals* (ADBI Discussion Paper, No. 54). Tokyo: Asian Development Bank Institute (ADBI).

Seufert, V., (2012). Organic agriculture as an opportunity for sustainable agricultural development. Available at: http://www.mcgill.ca/isid/files/isid/pb_2012_13_seufert.pdf.

Seufert, B., Ramankutty, N., & Foley, J. A. (2012). Comparing the yields of organic and conventional agriculture. *Nature, 485*(7397), 229–232. https://doi.org/10.1038/nature11069.

Shah, P. A., Brooks, D. R., Ashby, J. E., Perry, J. N., & Woiwood, I. P. (2003). Diversity and abundance of the coleopteran fauna from organic and conventional management systems in southern England. *Agricultural and Forest Entomology, 5*(1), 51–60.

Shiva, V. (1991). *The violence of the Green Revolution: Third world agriculture, ecology and politics: Ecological degradation and political conflict.* London: Zed Books.

Sirieix, L., Kledal, P. R., & Sulitang, T. (2011). Organic food consumers' trade-offs between local or imported, conventional or organic products: A qualitative study in Shanghai. *International Journal of Consumer Studies, 35*, 670–678. https://doi.org/10.1111/j.1470-6431.2010.00960.x.

Smith, A. (2006). Green niches in sustainable development: The case of organic food in the United Kingdom. *Environment and Planning C: Government and Society, 24*, 439–458.

Stamm, A. (2008). Agribusiness and poverty reduction: What can be learned from the value chain approach? In C. Stringer & R. Le Heron (Eds.), *Agri-food commodity chains and Globalising networks* (pp. 23–34). Ashgate: Hampshire.

Sylvander, B., & Kristensen, N. H. (2004). *Organic marketing initiatives in Europe*. Aberystwyth: School of Management and Business, University of Wales Aberystwyth.

Tallontire, A. (2000). Partnerships in fair trade: Reflections from a case study of Cafédirect. *Development in Practice, 10*(2), 166–177.

Thatiah, P. (2005). *Poison all over, but who really cares?* Nairobi: The Nation (Nairobi).

Thottathil, S. E. (2014). *India's organic farming revolution: What it means for our global food system.* Iowa City: University of Iowa Press.

Trouillot, M. R. (1994). Haiti's nightmare and the lessons of history. *NACLA Report on the Americas, 27*(4), 46–51. https://doi.org/10.1080/10714839.1994.11724636.

UNCTAD. (2008). *Trade and development report.* New York and Geneva: United Nations. Available at: https://unctad.org/system/files/official-document/tdr2008_en.pdf.

Van der Ploeg, J. D., Renting, H., Brunori, G., Knickel, K., Mannion, J., & Marsden, T. (2000). Rural development: From practices and policies towards theory. *Sociologia Ruralis, 40*(4), 391–408. https://doi.org/10.1111/1467-9523.00156.

Van Elzakker, B., & Eyhorn, F. (2010). *The organic business guide: Developing sustainable value chains with smallholders.* Bonn: IFOAM.

Van Hemelrijck, A. (2013). Powerful beyond measure? Measuring complex systemic change in collaborative settings. In J. Servaes (Ed.), *Sustainability, participation and culture in communication: Theory and praxis* (pp. 25–58). Intellect: Bristol.

Via Campesina. (2007). *Nyéléni Declaration Sélingué*. Mali: Forum for Food Sovereignty. Available from: http://www.foodandwaterwatch.org/world/global-trade/NyeleniDeclaration-en.pdf/view

Weis, T. (2007). *The global food economy. The battle for the future of farming*. London/New York: Zed Books.

Whatmore, S., & Clark, N. (2006). Good food: Ethical consumption and global change. In N. Clark, D. Massey, & P. Sarre (Eds.), *A world in the making* (pp. 363–412). Milton Keynes: The Open University.

Wilkins, K. G., & Enghel, F. (2013). The privatization of development through global communication industries: Living proof? *Media, Culture & Society, 35*(2), 165–181.

Willer, H. (2011, February 18). *Organic agriculture in Europe 2009: Production and market*. Switzerland/Nürnberg: Research Institute of Organic Agriculture FiBL/BioFach Congress.

Willer, H., & Kilcher, L. (Eds.). (2009). *The World of Organic Agriculture. Statistics and emerging trends 2009*. Bonn: IFOAM/FiBL, Frick.

Wilson, B. R., & Mutersbaugh, T. (2015). Fair trade certification, performance and practice. In L. T. Raynolds & E. A. Bennett (Eds.), *Handbook of research on fair trade* (pp. 281–297). Cheltenham: Edward Elgar Publishing Limited.

Wittman, H. (2011). Food sovereignty: A new rights framework for food and nature? *Environment and Society: Advances in Research, 2*(1), 87–105.

Woods, M. (2007). Engaging the global countryside: Globalization, hybridity and the reconstitution of rural place. *Progress in Human Geography, 31*(4), 485–507. https://doi.org/10.1177/0309132507079503.

World Bank. (2007). *The World Bank Annual Review*. Availabe at https://siteresources.worldbank.org/EXTANNREP2K7/Resources/English.pdf. Accessed 22 Oct 2015.

Zerbe, N. (2014). Jennifer Clapp: Food. *Agriculture and Human Values, 31*(1), 161–162.

CHAPTER 3

Sustainable Development Through Communication: Toward New Theoretical Approaches

Introduction

This chapter introduces communication as an analytical tool for the sociability of the global food economy and for a more nuanced understanding of sustainable development. Communication is believed to play a fundamental role in attaining sustainable development, not just through the sharing of information and knowledges but also by contributing to the interplay of different development factors (Servaes 2013). Although communication is not explicitly mentioned in any of the 17 SDGs, long before the publication of the Sustainable Development Goals, the intersection between sustainable development and communication has been the focus of attention of scholars in the field of communication studies (Servaes and Lie 2015; van de Fliert 2010, 2014).

In this chapter, the role of communication is approached primarily through the theory and practice of communication for development and social change (CDSC). Communication for development and social change involves the utilisation of a range of communication tools and media to achieve social change—environmental, political, or cultural—improve people's lives, and promote social justice. There are several reasons for taking this approach, starting with the explicit and systematic connection between communication and development that the field provides. Through a critical analysis of this connection, the field offers a fertile depository of conceptual tools and empirical data that can inform a more

M. Touri, *Sustainable Development and Communication in Global Food Networks*, https://doi.org/10.1007/978-3-030-46119-5_3

nuanced view of sustainable development. The critical work of communication and development scholars and practitioners has made inroads into our understanding of development as a multidimensional process and concept. More importantly, their work provides an entry point in the deconstruction of sustainable development through the communicative practices that reside in the social and economic relations that underpin the food market; and through the voices of the people as they emerge from these practices. By calling for communication to be studied in new contexts, the field also legitimises the need to look at the role of communication in development in global food networks.

Overall, the field has emerged out of various schools of thought that were developed by scholars and practitioners across the world, leading to the use of different names and discourses, and rendering the field a 'chameleon', embedded in international development; changing colours to reflect the development thinking of the day (Quarry and Ramírez 2009: 6). Among the most dominant discourses are the following: development communication (DevCom), communication for development (ComDev), Latin American alternative communication, and communication for social change (CFSC). Communication for development and social change has also been practiced under the names of Information, Communication and Technology for Development (ICT4D) and Learning for Social Change, and Social and Behavioural Communication (Thomas and van de Fliert 2014; Tufte 2017).

More recently, the name Communication for Sustainable Development and Social Change (CSSC) has also been added capturing the multidimensional model of sustainable development and the need for a multi-stakeholder approach to policymaking in which communication plays a critical role (Servaes 2013: 5–6). In this book, the name Communication for Sustainable Development and Social Change (CSSC) is used as a term that encompasses the premises of the different approaches. More importantly, CSSC provides a more accurate reflection of the book's effort to contribute to the deconstruction of sustainable development through communication and through a critical analysis of development.

The chapter begins with a brief review of the two dominant paradigms that have defined the field of CSSC, namely the modernisation and participatory paradigm. This review paves the way for a discussion on the role of communication in rural development and agriculture, which is the broader context for food production and food networks. This is followed by a critical analysis of the economistic discourse that has shaped our understanding of communication in sustainable development. Communication scholars articulate the need for new concepts to think with, new contexts, stakeholders, and dynamics that will provide new opportunities for innovative pathways to development and will promote the voice and agency of marginalised groups and communities. Through this critique, the field provides a legitimate analysis of the importance of communication theory and practice in understanding and attaining sustainable development.

Drawing on some of the key lessons learnt from the field of CSSC, the chapter employs the language of weak theory to introduce the concept of *weak communication*, a concept that aims to capture the essence of the less visible social relations and modes of communication, and their link to sustainable development in new contexts. The concept is built on theoretical and empirical insights from the field of CSSC and from theories and concepts from the broader discipline of communication studies. Through this concept, the chapter calls for a rethinking of the connections between communication, sustainable development, and social change and for a renewed focus on communication practices that fall outside the boundaries of institutional development programmes and of dominant theories and models of communication, development, and well-being more broadly.

Communication for Sustainable Development and Social Change: A Brief Review

Over the years, the study of CSSC has witnessed several paradigmatic shifts in theories and models of economic and social development, but two schools of thought have dominated the theoretical and empirical orientations in the field: the modernisation paradigm and the participatory paradigm. In his recent book *Communication and Social Change: A citizen's perspective*, Tufte (2017) analyses the field through the prism of a three-generation model based on different generations of experiences. The first generation was concerned with the diffusion of innovations and dissemination of information that focused on individual behavioural change, the second generation focused more on educational communication and the development of people's own skills and capacity to handle their lives' challenges, and the third generation finds expression in the communication for social change approach that emerged from Brazilian educator Paulo

Freire's (2001) methods and seminal work that champions the idea that people have the ability to analyse their own reality and find appropriate solutions for their needs and local conditions (Tufte 2017: 16). Melkote and Steeves (2015) approach the field through two major schools that they label as behavioural change communication and communication for social change, reflecting the modernisation and participatory paradigms, respectively. This binary broadly reflects the dichotomy between DevCom and Communication for Social Change (CSC) models where the former is guided by the principles of modernisation and the latter agrees with the participatory communication approaches (Melkote and Steeves 2015: 381–382).

The Modernisation Approach

In line with theories and discourses that see development through the prism of economic growth, the modernisation approach adopts a very linear and top-down process of communication that is based on the creation of effective messages targeting audiences with the aim of persuading them to change their behaviour. This was indeed one of the most powerful paradigms originating after WWII, based on which West European and North American countries were used as models of development that should inspire and be replicated by the countries in the Global South, with the assistance of communication practices. Modernisation approaches viewed communication as a reinforcer of development, primarily through the role played by mass media and their pro-development content, in the countries of the Global South. The underlying assumption was that through the dissemination of innovative and progressive ideas, a climate of modernisation would be encouraged, leading to the desired behavioural change among rural populations and the removal of traditional attitudes. Several critiques pointed to the numerous flaws and limitations of this approach. Not only was the disseminated media content not pro-development, but it was most likely designed to promote the political and economic interests of the local elites. Moreover, economic and political inequalities in the countries of Asia, Africa, and Latin America meant that any innovations that were adopted by the populations were limited to the more privileged and educated groups, resulting in even greater inequalities (see Melkote and Steeves 2015). Consequently, it is unsurprising that several of the programmes that followed the guidelines of the modernisation approach failed.

The Participatory Paradigm

The participatory paradigm draws from Paulo Freire, the Brazilian philosopher who conceived of communication as dialogue and participation, with a strong emphasis on the recognition of non-mediated and interpersonal communication, dialogue, and persuasion, where communication is a cyclical process of information sharing, and leads to individual and collective change (Rodriguez 2011; Thomas 2011). Participatory development communication involves the participatory use of different forms of media and communication methods towards development. Such media and methods include community radio, ICTs, or interpersonal communication. These activities enable meaningful knowledge generation and exchange, skill development, and establishment of a platform for collective decision-making and action (Thomas and van de Fliert 2014). At the heart of this process lies the important role of people and the alternative knowledge they could generate, which should be acknowledged and respected. In the participatory framework, communication is approached as a meaning-making process that respects and emphasises the cultural contexts and processes through which communities understand and act upon their lives (Dutta 2011). Therefore, what matters is not just the availability of information to facilitate change among marginalised groups but also the quality of relationships between speakers and listeners. Here, the focus is on the empowerment of communities through their active participation in the identification of problems and solutions and through a process of collective action and reflection (Tufte 2017).

Although communication is an indispensable component of participatory approaches, the role of the media changes. Mass media are no longer considered sufficient as the attention is shifted to different communication strategies that are designed in response to the needs identified by the communities themselves. In this case, the communication role of development agencies is expected to be less prescriptive than under the modernisation approach (Melkote and Steeves 2015). Moreover, although most participatory communication programmes are run by agencies and institutions, participatory communication approaches take place also outside institutions through community-based communication practices (Tufte and Mefalopoulos 2009). These are cases of culture-oriented collective action, grassroots movements, and citizen participation in the public sphere (Tufte 2017: 76–79).

Communication in Agriculture and Rural Development: Opportunities and Challenges

Several of the communication strategies that have been used in rural development and agriculture fall under the two dominant schools of thought. On the one hand, and following the logic of information diffusion, communication campaigns and marketing techniques are used to raise awareness and influence the attitudes of rural populations on environmental issues. On the other, scholars and practitioners that work in the participatory school of thought tend to use more creative and participatory methods such as storytelling and participatory video (Servaes and Lie 2015). In the area of food and agriculture more specifically, communication efforts have focused on the production and sharing of information and the diffusion of agricultural technologies, and on the use of participatory communication processes with the involvement of local communities and local knowledge in the identification of problems and solutions. Since the 1970s, communication has been a vital component in the diffusion of information about agricultural innovations, with the mass media playing an instrumental role as information providers and facilitators in changing traditional behaviours and practices in the Global South (Dutta 2011). There is now a wealth of studies on communication interventions facilitated by international organisations and local NGOs, which aim to assist farmers to acquire the skills and technologies that will enable them to increase their food production.

In the last three decades, participatory communication approaches have also become an essential component in development projects for achieving economic, social, and environmental sustainability. Participation has therefore been recognised as key in empowering communities and individuals to determine their own development objectives and make informed decisions (Van de Fliert 2007, 2010). A leading figure in the introduction of participatory approaches in agriculture and food production was the British development scholar and practitioner Robert Chambers, a pioneer in the design of alternative and participatory methods of development. In his 1983 book *Rural Development—Putting the First Last*, Chambers offers a detailed and rich with examples analysis of the ways in which research and knowledge on rural development is dominated by centre-periphery biases that fails to recognise the knowledge of local people and marginalises anything that is indigenous, human, and organic (Chambers 1983: 77). Inspired by Paulo Freire's seminal work *Pedagogy of the*

Oppressed (1970), Chambers describes experts as 'outsiders' who fail to learn from rural people's knowledge due to numerous forces including power, prejudice, and lack of contact, as well as an unfortunate communication gap between academics and practitioners who are willing to spend time with and learn from indigenous knowledge (ibid: 83–84). It was also through Chambers' work that the idea and process of participation became popularised in the study and practice of development. His Participatory Rural Appraisal (PRA) approach (Chambers 1994) promoted the use of carefully designed methods to facilitate the process of analysis by rural people, the need for active listening on behalf of experts and outsiders, and the need for collaboration and knowledge sharing between local rural communities and experts. At the heart of this methodology lies the belief that communities have the right to be included in projects that are set up for their benefit, regardless of their level of formal education or literacy skills (Melkote and Steeves 2015: 378–9). The PRA method has been used successfully in many parts of Africa with the employment of group community work where communities have taken the lead in planning their development action and identifying solutions, while researchers have mostly served as facilitators (Anyaegbunam et al. 2004). In fact, Chambers' work (1983, 1994) provides evidence of the value of local people's intimate knowledge about the environment and the soil, but such knowledge was considered as backward and unscientific by outside experts. Once again, Chambers is openly critical towards scientists and officials, who

> ...believed that their knowledge was superior and that the knowledge of farmers and other local people was inferior; and that they could appraise and analyze but poor people could not ... Treated as incapable, poor people behaved as incapable, reflecting the beliefs of the powerful, and hiding their capabilities even from themselves. Nor did many outsider professionals know how to enable local people to express, share and extend their knowledge. The ignorance and inabilities of rural people were then not just an illusion; they were an artifact of outsiders' behavior and attitudes, of their arrogant and ignorant manner of interacting with local people. (Chambers 1994: 963)

Chambers himself acknowledges that participatory methodologies and PRA approaches are not panaceas, but they still create opportunities for the voices of the people normally excluded to be heard and to influence policymaking in agricultural development (Chambers 1997).

Communication interventions for farmers' training are generally geared towards more participatory involvement of the farmers, for defining their own problems and actions, as well as the information and skills required to carry out these actions (Coldevin 2003; FAO 2015). Participatory methodologies have been widely used by the FAO Communication for Development Group, which is the leading practitioner of applied communication for agricultural development for the last 35 years. It has a rich record of programmes and reports on the use of communication practices that have focused on knowledge acquisition, attitude change and confidence-building, participation in decision-making, and action, all required processes of education and communication (Coldevin 2003). The FAO has incorporated communication for development into its practice using media and participatory tools (Balit 2012). It uses the term 'ComDev' which it defines as social process that is based on dialogue and integrates the media with local communication systems to increase the involvement of rural populations in development initiatives and facilitate active participation (FAO 2014). A specialised FAO ComDev Group is also part of the Research and Extension Branch within the Office of Knowledge Exchange, Research and Extension (Thomas and van de Fliert 2014: 22). The FAO has also used the so-called Participatory Rural Communication Appraisal model (PRCA), a model that is similar to PRA but with emphasis on communication issues and methods. It is a method that encourages collaboration between local communities and external experts and aims at facilitating collaborative knowledge generation and design of communication processes (Melkote and Steeves 2015: 393).

However, good participation is difficult to achieve, and participatory approaches are more likely to end up being used to increase the effectiveness of an externally introduced programme that emphasises economic growth, rather than empower people to define and take control of their own development and well-being (Van De Fliert 2010). Cases like FAO's agricultural project PRODERITH[1] in Mexico show that in large-scale and

[1] PRODERITH (Programme of Integrated Rural Development in the Tropical Wetlands Project) was a program funded by the World Bank that run in two phases. Phase 1 ran from 1978 to 1984 and focused on improving agricultural development in Mexico's wetlands, achieving participation of all involved, including peasants and PRODERITH staff, and assume an integrated rural development nature. The second phase ran from 1986 to 1995 and was committed to promoting community members as protagonists in their own development. Through training workshops for community members, it aimed to create spaces where communities could gather and discuss community issues. Communities also received training

long-term programmes, the communication and participation aspect is treated in an instrumental way and tends to get lost in the end, with communities being unable to be involved in their own development (see Fraser and Restrepo-Estrada 1998; Melkote and Steeves 2015). Very few development initiatives manage to go beyond consultative participation and only when the agencies involved build long-lasting collaborative relationships with the communities. In most cases, so-called participatory approaches rely on tools and methods that are externally designed and demonstrate limited understanding of the underlying principles of development (Van De Fliert 2010).

A similar lack of understanding of the complexity of achieving development through communication can be observed in the employment of information and communication technologies (ICTs). Without negating the value of ICTs in development, their use often fails to address the real needs of communities. Since the 1990s, the FAO has worked with internet-based ICTs, starting with the establishment of farmer-operated networks in Latin America. These networks consist of internet-based linkages between producers, farmer associations, and NGOs. In these networks, ICTs and conventional media are believed to contribute to integrated development through various uses and applications, including finding markers for produce and sharing information on markets prices and consumption trends. The internet can facilitate farmers' access to global markets and information about consumer needs and realities (Melkote and Steeves 2015: 291–292). Such information can prove valuable in assisting farmers to plan for future crops and increase profitability. A successful example of such a network was the Farmer Information Network (FarmNet) that was developed by FAO along with the Uganda National Farmers' Association (UNFA) (Crowder 2000).

As the access and usage of mobile phones among rural communities increases, mobile phone-based services are also becoming more popular with farmers, thanks to the ability to send agricultural information as a text or recorded message, through interactive voice response systems, and mobile applications. Text-based messages can be particularly useful as they can be customised, they can be provided in the local language, and they can also be stored on the device (FAO 2013). In fact, thanks to the wider availability of cheaper mobile phones, the lower network charges, and the

in concepts of rural development and in the production of video and printed materials (see Fraser and Restrepo-Estrada 1998).

newer applications, mobile phones are the most frequently used ICT for access to information about market prices, volumes, and weather-related information that can help farmers plan their production and investments. An example of an affordable SMS-based service is the Reuter's RML Information Service that was launched in India in 2007, providing subscribing farmers with daily updates on prices, markets, and weather (FAO 2017: 4). Mobile phones have been used successfully in several other programmes such as the African Cashew Initiative, the Lifelong Learning for Farmers programme in Uganda, and the iCow initiative in Kenya. These programmes provided interactive SMS services that enabled farmers to get access to information about prices, storage practices, and other agricultural innovations (see Sousa et al. 2019). Other information-based solutions also include the public access points such as tele-kiosks and telecentres, hardware solutions, and e-governance and connectivity (Thomas and van de Fliert 2014: 88).

Yet, despite the numerous commendable initiatives, the use of ICTs in rural development offers limited opportunity for farmer participation, failing to achieve local input and local ownership (Sulaiman et al. 2012). The introduction and application of ICTs by government officials, international agencies, and NGOs has followed a linear approach that focuses on a narrow and universal transfer of technology and information dissemination (Keniston and Kumar 2003). ICTs continue to remain exclusionary, especially among farmers who lack the required education and ICT literacy or even the time to learn how to operate modern technologies. In fact, technological progress and the availability of Information and Communication Technologies (ICTs) has led to the design of development projects and interventions that are driven by resource supply rather than demand without considering the needs of the people (Manalo and van de Fliert 2011; Unwin 2009). This is even though traditional media such as community radio remain particularly popular among agricultural communities mainly due to its inexpensiveness and accessibility, especially among poor communities with low literacy skills. It is also an effective platform for sharing indigenous knowledge around food preservation methods, crop varieties, and planting and harvesting times (FAO 2013). As demonstrated in the study by Manalo et al. (2009), farmers in the Philippines preferred print publications as an information delivery method during agricultural training. The use of group media methods such as slide presentations, videos, audiocassettes, and theatre have also proved to be more popular in FAO rural development projects, especially as a means for agricultural extension, that is, the application of scientific research to agricultural practices through farmer education. The use of video enables a more effective communication between farmers and extension systems through information that is packaged in easily understandable ways. Digital Green in India is a well-known example of an organisation that has harnessed the visual power of video. The organisation focuses on training farmers to make their own videos to communicate and share their problems, solutions, and success stories to boost farm productivity and improve nutrition (Gandhi et al. 2009). Video has been found to facilitate learning through sharing of high-quality information among rural populations. One of its strengths is its compatibility with other media such as radio, TV, and mobile phones (Bentley et al. 2014; Chowdhury et al. 2015; Van Mele et al. 2010). In fact, the use of video on mobile phones is believed to be a promising solution, which has already been used for information sharing regarding agricultural practices and innovations (Baributsa et al. 2010; Bentley et al. 2014).

Examples like the above show the importance of a convergence of more traditional, easier to use media, and more advanced technologies, but they also highlight the need to separate the role of technology from the information it delivers, the choices it creates for people, and the constraints that it poses to those who lack the necessary resources to benefit from them (Heeks 2010). The real success of these ICT solutions depends on socio-economic factors such as age, gender, and income that continue to define the intensity of use of ICT tools, as a source of agricultural information for smallholder farmers, with large parts of the rural population in Africa, Latin America, and India remaining unreached; while challenges facing the use of ICT in dissemination of agricultural information may even contribute to further inequality in agricultural productivity and economic performance (Musa et al. 2013; Tata and McNamara 2016). Many of these projects have proved unsustainable after the donor funding ends due to approaching physical access to technology as a development goal, while failing to equip communities with the skills and technical support they need to maintain the project (Hosman and Fife 2008). What needs to be considered is that the transformative effect of ICTs on people's lives is highly personalised, and technology might support outcomes that go beyond economic growth. This became evident in Dorothea Kleine's (2013) study on the role of ICTs in people's lives in Chile, where telecentres, mobile phones, and internet-based technologies allowed people to make choices and achieve outcomes of non-economic nature. In the context of the food market, Sarah Lyon's (2018) study of coffee growers in Mexico also reveals the political activism dimension of development afforded by digital media use.

Critiques of the Field and the Way Forward

The employment of communication practices and technologies in the area of food production and agricultural development is reflective of the economistic discourse that has driven development initiatives in agriculture and food production and has also shaped our broader understanding of sustainable development with a focus on replicating the economic growth and modernity of advanced economies across the globe (Chap. 2). For the most part, the literature and existing studies on communication and ICTs in rural development favour a much econocentric and market-focused approach, which encompasses the neoliberalism and capitalism focus of communication and development agendas. Implicit in this discourse is an intellectual approach to sustainable development and social change that is built on Western ideas of development and neglects the array of innovative development and communication practices that have taken place since the 1930s and embody alternative understandings of development that divert from Western ideologies (see Manyozo 2012; Melkote and Steeves 2015; Tufte 2017). This is despite the emergence of new paradigms of development that have shaped post-development discourses and that range from China's own growth model to models of sustainable development that are inspired by the Latin American concept of Buen Vivir as a possible alternative to development (Chassagne 2018; Tufte 2017). There is now a wealth of critical development studies that interrogate not just the theory and practice of communication in achieving development but the actual meaning of development per se (Fair and Shah 1997; Melkote 2003; Dutta 2011; Wilkins and Enghel 2013; Thomas 2014, 2015; Wilkins 2000). These critiques converge in the institutionalisation of participatory communication approaches and the inadequacy of the dominant results-oriented, institutionalised approaches that fail to consider the importance of local contexts and power relations in development interventions. Instead they prioritise modernisation as 'a neoliberal project that legitimizes free markets through the privileging of individual empowerment as an approach to social change' (Wilkins and Enghel 2013: 166). It appears that what Robin Mansell critiqued in 1982 as a preoccupation of research and practice in this area with quantification and measurement of 'external' benefits (Mansell 1982: 48) continues to characterise development initiatives, undermining the contextual nature of development and the need for development to be understood in line with a particular culture and people (Makuwira 2018). What is even more worrying is the way in which the prescriptive and techno-centric approach of many communication projects and programmes has persisted over the years both empirically and theoretically. This was highlighted by practitioners Wendy Quarry and Ricardo Ramírez in their 2009 book, *Communication for Another Development: Listening without telling*, where they make a compelling case about the need for more small-scale projects working at a regional and local level, where face-to-face communication and 'listening' becomes an integral and essential component of the development progress. However, as the authors attest, not only is the development industry approaching communication as a bureaucratic process that first meets their public relations needs, but it is also unwilling to learn from past mistakes, and 'somehow the listeners lost ground to the tellers' (Quarry and Ramírez 2009: 34).

Thomas and van de Fliert (2014) are critical of the omission of power, especially in academic discussions, which they attribute to the institutionalisation and continuing impasse of 'participation', since the introduction of participatory models has not really diminished the prevalence of the diffusion model. 'In fact, the language of participation has been co-opted and eviscerated of the original meanings that Freire and others had given it' (Thomas and van de Fliert 2014: 11). The participation processes implanted into large-scale projects funded by organisations bodies such as the World Bank, the United States Agency for International Development (USAID), and Department for International Development (DFID) have brought questionable results based on vaguely articulated principles and indicators of development and social change. More importantly, it is the economic and political agendas of these institutions that determine how participation is implemented, often imposing 'participation' on people and setting the programme up for failure. Participation should be a choice and should be in line with the community's socio-cultural context and values; but implementing the participation paradigm in the field remains a challenge for practitioners. In many so-called bottom-up development and social marketing programmes where people were invited to participate in self-help activities, in reality, they were manipulated to take part in projects that would make them dependent on the global market as consumers of specific products and services (Melkote and Steeves 2015; Ramalingam 2014).

Finally, Tufte (2017) summarises these challenges in a review of the main traits of the different paradigms and generations of the field. These traits amount largely to an institutionalised practice of communication, which is constituted through the narrow focus on pre-established and tangible quantifiable targets that can be achieved through the transfer of information, and the employment of strategic communication methods and approaches. The result is the overshadowing of the dynamic, open-ended, and often intangible processes and relations that are inherent in empowerment and development. Instead, communication interventions tend to remain focused on changing behaviours rather than bringing more meaningful social change. Tufte's critique seeks to urge scholars to take a step back from analysing communication strategies implemented by institutions and development agents. Instead, he proposes a 'citizen perspective' to the study of communication for sustainable development and social change and calls for a less instrumental approach and understanding of the role of communication, one that is based on new contexts, stakeholders, and dynamics (Tufte 2017: 31–32). New contexts are already emerging through digital media developments, global citizen insurgencies, and practices of bottom-up communication and collective action. These are liminal spaces that celebrate the agency of oppressed groups, they allow for genuine exploration of deliberative development alongside communities (Manyozo 2017: 3). They provide many and new opportunities for innovative pathways to development and, more importantly, for new pathways to re-write the intellectual history of development through the theories and practices that extend beyond Western discourses (Tufte 2017: 31–32).

But in order to unearth, explore, and learn from these new contexts and the development opportunities they encapsulate, the knowledge and voices of the people and of marginalised communities need to be heard. The work of Indian media scholar Mohan Dutta offers a valuable intellectual ground as he advocates for the voice and visibility of marginalised communities. In his 2011 book *Communicating Social Change: Culture, Structure and Agency*, he argues that it is only by writing history from below based on dialogue and listening to the voices of the marginalised populations that discourses of development can also be re-written (2011: 7). The transformative power of this dialogue constitutes the core of his Culture-Centred Approach (CCA) in which Dutta promotes a connection between culture, structure, and agency, where culture is 'defined in terms of the continuously shifting local contexts and frameworks of meanings'

(Dutta 2011: 174). The theory aims to interrogate the notions of development through the presence of subaltern voices in making claims of development (Dutta 2015). At the heart of the CCA lies the transformative role that dialogue can play in changing the structures of the status quo and combating the injustices perpetuated by these structures. Dutta (2011) refers to the structures of neoliberal development models and the West-centric production of knowledge, where the voices of marginalised communities have been systematically erased due to the inequities in the distribution of communication platforms. Hence, a dialogic approach to development and social change can offer a way to reverse the unequal distribution of communication resources allowing the articulation of development problems from a subaltern standpoint. Such an approach requires reflexivity and listening to subaltern voices and engagement with the marginalising social structures that need to change.

Departing from Dutta's Culture-Centred Approach and building on Chamber's proposition about the power of language in development (Chambers 1997), the idea put forward in this book is that in order to unearth opportunities for marginalised communities to articulate their voices and contribute to changing the structures of the status quo in the theory and practice of sustainable development, a change of language is paramount. If language influences agendas and mindsets and stimulates research and learning around development (Chambers 1997: 1744), we need to bring forward and create space for the language and mindsets of marginalised groups and alternative disciplines. This idea finds expression also in the proposition that knowledge is performative and social scientists actively participate in constituting the object of study (see also Law and Urry 2004). In this book, inspiration for a new language is found in two areas: the weak theory introduced in Chap. 2 and the marginal, informal, and non-institutional forms of communication, social relations, and development that have remained understudied and undertheorised.

Towards a Weak Theory of Sustainable Development and Communication: The Power of *Weak Communication*

Chapter 2 introduced weak theory, inspired by Eve Sedgwick's work, as an analytical approach that can offer richer, more focused, and localised descriptions and explanations. In this chapter, the theory provides the grounds for developing a new concept, the concept of *weak communication*.[2] This is a new concept to think with and to analyse the role of communication in sustainable development. In fact, Sedgwick's criticism to strong theory and her call for more attention to the rich description of a moment resonates with critiques towards dominant and institutional approaches to sustainable development and social change. Easterly's (2006) division of the development industry into 'planners' and 'searchers'—where the former set big goals and adopt a top-down 'transformational' approach to development, but neglect the complexity of ground reality, while the latter are process-oriented who pay attention and listen to the people on the ground—reflects a similar logic and division between strong theory and its opposite, weak theory. Perceptions of 'development' as the identification of problems, planning, and implementation of solutions that will bring social, economic, and political transformation in low-income countries embody a framing of development that is as reductionist as any strong theory that pays very little attention to local contexts of well-being (Makuwira 2018). Inherent in these strong theory-oriented approaches to development is an equally reductionist approach to communication that neglects the legitimacy of more marginal communication practices and the different types of well-being that these can enable. I propose that a new theorisation for all those less visible practices and their role in enacting equally less visible forms of sustainable development is encapsulated in the concept and power of *weak communication* (see also Touri 2020).

The concept of weak communication captures the power of the 'silent' and less visible relations and communicative practices that support larger-scale aspects of social structure and social change. These practices are weak in the sense that they are less 'noisy' (Tufte 2017) than the more mainstream types of communication that sustain social movements and social change campaigns. They are also weak in terms of lacking the institutionalised, scalable, and quantifiable character of communication practices that

[2] The concept of weak communication was developed during a conceptual Hackathon held at Loughborough University London, in September 2017, with a global research network of scholars working in this field. The Hackathon led to an edited collection of innovative concepts related to the role of communication in processes of change. Inspiration for the concept of weak communication was also drawn from the work of economic geographers Katherine Gibson and Julie Graham on diverse economies and the weak economy, which was discussed in Chap. 2. For a more extended description of the concept of weak communication, see Touri (2020).

are meant to enable development. Yet, they remain powerful in their capacity to support and enable social change in subtle but significant ways. A typical example of weak communication is the interpersonal and dialogic communication through which people can give and receive information. Interpersonal communication enables meaningful knowledge generation and exchange, skill development, and establishment of a platform for collective decision-making and action (Rodriguez 2011; Thomas and van de Fliert 2014). Yet, it remains neglected and undertheorised.

Paulo Freire's (2001) revolutionary pedagogy has paved the way for an understanding of dialogue as an empowering process of liberation from the unjust structures of society. Freire saw interpersonal communication between teacher and student as a process of exchange that empowers people to become aware of their own oppression, elevating the role of interpersonal communication in social change more broadly. The power of direct and interpersonal communication has, of course, been recognised in the context of communication for development programmes where community members could identify problems and suggest their own action strategies. What have received less recognition are the dialogic processes that are embedded in people's daily activities and can facilitate development and empowerment in less visible and unforced ways. Informal communication channels such as through friends, family, peers, markets, and festival gatherings provide spaces where people not only give and receive information but can also inform their decision-making and action in ways that can lead to more permanent change (McAnany 1980). In these channels, the 'listening' side of the communication is critical, bringing into visibility the heterogeneity of actors and practices that can contribute to and constitute development. Listening embodies recognition of indigenous knowledge in development, even though the empowering aspect of knowledge sharing has traditionally been the property of the expert and the scientist (White 1994). Yet, listening has lost ground and credibility, and despite being 'the function of communication that nurtures our self-awareness and sense of confidence, the one that helps us change our situation in our own terms' (Quarry and Ramírez 2009: 20), it falls through the cracks of strong theoretical readings of communication in development and social change.

Practices such as the simple act of listening for development, and specifically for development that happens outside the formal boundaries of institutionally driven programmes, are 'weak' because they open new avenues to surprise and understanding of the world. They also encourage a different approach to development, one that is not tied to economic growth and income generation for marginalised communities, but one that recognises the multiple dimensions of well-being and capability building. These include less measurable and quantifiable aspects of development and growth, including communities' self-awareness, confidence, and motivation to work towards achieving social change in their own communities. In this sense, weak communication may be difficult to replicate or make scalable, but it does not need to expand its reach and reductiveness. It is a call for scholars to pay more attention and listen to the people on the ground without being driven by continuous suspicion and fear of surprise.

A note of clarification is needed about what weak communication is *not*. Weak communication is not meant to provide a universal approach to interpersonal and dialogic communication as this would cancel out its 'weak theory' underpinning. Nor is it meant to theorise and define a specific type of communication. Instead, it is an attempt to reconceptualise how we approach communication, development, and social change, and expand and enrich scholarly and practical work in the field (Tacchi and Tufte 2020). Weak communication is about being open to new, emerging connections between communication, sustainable development, and social change that fall outside the boundaries of institutional development programmes and of dominant theories and models of communication, sustainable development, and well-being more broadly. Practices such as dialogue and listening represent typical examples of communication that fall outside those institutional boundaries. From another perspective, weak communication is also about being open to ideas and insights from different, more localised real-life contexts, and from different fields of communication research, including interpersonal, organisational, and small-group communication research, and from the political and social sciences more broadly. The following sections present a discussion of theories and concepts that are fruitful in demonstrating the different types of communication that happen outside institutionalised development programmes and can still contribute to community well-being and sustainable development. Although the discussion that follows is by no means complete in its consideration of relevant theories, it is indicative of how we can create entry points for new knowledge in the field of communication for sustainable development and social change.

At this point, it is imperative to also add a note of clarification regarding the relation between the concept of weak communication and Mark

Granovetter's seminal theory of weak ties, with which readers might naturally draw parallels. Granovetter (1973) distinguishes between strong and weak social ties that characterise social and economic networks. The strength of these ties depends on emotional intensity, intimacy, and trust. According to the theory, the weak ties are most valuable for information and knowledge diffusion across a network. The notion of weak communication introduced here should not be confused with Granovetter's weak ties. Weak tie theory is more concerned with networks and the strength of relationships. Instead, weak communication aims to capture the value of less formal types of communication, regardless of the strength of relationships through which this communication takes place. Moreover, weak communication is a concept to think with, and to think more critically about the connection between communication, sustainable development, and social change. Ultimately, the concept of weak communication offers a different choice in the kind of knowledge that we produce encouraging a performative orientation to knowledge. This is the knowledge that stems from a thick description of communication practices that directs attention to the nuances, affects, and multiple codes of meaning in communication.

Interpersonal Communication

In the previous section, interpersonal communication was introduced as one of the typical examples of the less noisy types of communication that, despite being instrumental in people's well-being, remains less studied and theorised in the context of development. The role of interpersonal communication has been underlined by studies dating back to Lazarsfeld's two-step flow process (Lazarsfeld et al. 1944), particularly with regard to the ways in which mass media interventions could stimulate interpersonal communication processes that would lead to further behavioural change (Morris 2003). Interpersonal communication scholars can offer additional insight into the various routes through which this type of informal communication can foster development and well-being. Looking at interpersonal communication through the multiple goals theory, Dillard and Wilson (2014) approach interpersonal communication as purposeful and argue that it can serve multiple social goals—instrumental and relational—including overcoming obstacles, staying safe, and maintaining valued relationships. According to Solomon and Vangelisti (2010), the development of valued relationships is the most basic and consequential as relationships become a valuable source not just of companionship and support but also of information and knowledge. Drawing on social exchange theories, the authors explain how individuals tend to enter relationships to gain access to tangible or intangible resources and to maximise rewards and minimise costs that are naturally linked to their well-being. The connection between interpersonal communication and well-being has also been theorised through the idea of social support, which is defined as verbal and non-verbal communication between recipients and providers that reduces uncertainty about the situation, the self, the other, or the relationship, and functions to enhance a perception of personal control in one's life experience (Albrecht and Adelman 1987: 19). In this case, the contribution of interpersonal communication to well-being is understood to take place through the strong connections between social support and positive psychological and physical health outcomes (Cunningham and Barbee 2000; Ikeda and Kawachi 2010). Such outcomes are often linked to the feelings of connection that individuals experience through communication with other members of their social network. Interpersonal communication scholarship points towards the significance of social relations and interactions for well-being. This is a terrain that has been covered in-depth through the concept and theory of social capital, which can provide another fruitful analytical concept, helping draw more nuanced connections between communication, development, and well-being.

Social Capital

As a concept, social capital originates from sociology, political science, and economics and assumes the importance of and investment in social relations and practices of engagement for securing specific benefits for societies to function effectively. Social capital has been used widely in communication scholarship and research (Kikuchi & Coleman 2012). However, it does not have a universally agreed definition and meaning, with different studies adopting definitions according to the respective discipline (Robison et al. 2002). Despite the lack of consensus in terms of its conceptual and operational definition (Lee and Sohn 2016), there are certain theoretical and empirical dimensions that remain prominent.

Broadly speaking, social capital is premised on the ability of individuals and groups to use social relations to cooperate and act for common purpose and benefit. Approaches and definitions of social capital can be categorised based on their focus on (a) 'bonding' relations between individual actors, and (b) the structure of 'bridging' relations within a collectivity (Adler and Kwon 2002). While the former category emphasises the benefits of community groups (Coleman 1990; Putnam 1993), the latter approaches benefits as specific to the individual and embedded within relationships with other individuals or organisations (Bourdieu 1986; Burt 1992). Sodano and Verneau (2006) distinguish between two different strands of literature through which social capital has been theorised. Based on the work of Putnam (1993) and Fukuyama (1995), the first stand approaches social capital as the sum of social relations and networks that contribute to civil society, democracy, and economic development by facilitating social and economic cooperation as opposed to free-riding. One of Putnam's (2001) main observations is that social capital fosters cooperation and reciprocity among individuals in their pursuit of common goals. This approach emphasises the importance of trust and reciprocity among social actors, which is considered a key element for encouraging strong interpersonal relationships and collaboration, and can be seen as a precondition and an outcome of social capital (Fisher 2013). According to Putnam (2001), trustworthiness lubricates social life, while Reimer et al. (2008) approach trust as a matter of shared perspectives (see Rivera et al. 2018: 76). The second strand of literature draws on the work of Burt (1992) and Lin (1999) where social capital is approached as the 'resources embedded in a social structure that are accessed and/or mobilized in purposive actions' (Lin 2001: 29). Such resources could include an actor's contacts and the information and opportunities that can be reaped from these contacts. According to Sodano and Verneau (2006: 198), this approach emphasises the competition—rather than the cooperation—that is encouraged by networks, relations, and resources. Yet, a common thread that cuts across different approaches is the focus on relations that bring productive benefits, and such relations would be impossible without communication.

More importantly, social capital has been used as an analytical concept in poverty alleviation and rural development. Drawing on Putnam's (1993) definition of social capital as a set of interconnected networks underpinned by shared norms and beliefs, the World Bank describes social capital as 'critical for poverty alleviation and sustainable human and economic development' (2001). The relevance of social capital for rural development is premised on the way it affects people's capacity to get things done collectively and to organise themselves (Portes and Landolt 2000). In fact, the World Bank has recognised the impact of social capital on poverty, and in the Panama Poverty Assessment (2000: 22–23), it concluded that government and NGO assistance was more likely to reach rural communities with higher levels of social capital. In several countries in South America and the African region, evidence suggested that social capital was associated with collective action, which led to neighbourhood development, increased household welfare, and the ability to cope with economic adversity. Moreover, in a review of rural development projects around the world, Khan et al. (2007) have presented several examples of rural development facilitated by social capital. These examples included biodiversity and resource management, land conservation, and fertiliser adoption. In most cases, positive development outcomes were achieved through the diffusion of knowledge and new technologies and the spread of new ideas. This was facilitated by social capital practices including participation in community groups and sharing of ideas and experiences, rendering social capital directly relevant to rural development challenges. The role of social capital in development could also be understood through its analytical attention to social and cultural factors such as solidarity, sociability, and aspirations and the underlying shared values that affect economic transactions and outcomes. As research on rural African economies has demonstrated, access to productive resources and economic relations depends on participation in a variety of social relations and informal networks (Fafchamps and Minten 1999, 2002; GERGS 2003). Although we are reminded that economic relations are also underpinned by power asymmetries and can have negative social and economic effects, the value of social capital for understanding economic relations is that it draws attention to the ways in which loosely connected people rely on reciprocity of trust to perform their economic transactions. Through this trust, social relations can lead to a variety of economic benefits such as reducing transaction costs, enabling collective action, and generating new knowledge (Van Staveren and Knorringa 2007).

The link between social capital and development processes offers a useful conceptualisation of the role that relations and the communication processes that underpin those relations can contribute to development outcomes outside institutional settings. Here, communication involves social relations, interpersonal communication, and information gathering, exchange, and interpretation, which in turn facilitates community cohesion and collaboration, identification of problems and solutions, and can lead to positive social change. Ultimately, social capital becomes another useful conceptual tool that can account for the relationship between different types of communication and development in various settings of everyday life and among different actors, without the intervention of development agencies. The interplay between these different communication and development connections can be expounded further through knowledge offered by communication scholars in the field of applied and organisational communication. Through key concepts such as dialogue, voice, and community collaboration, this field can offer further conceptual clarity into the different connections that can be drawn between communication, development, and social change.

Communication Through Community Collaboration

In the field of applied communication and organisational communication, attention has been paid to the significance of dialogue as a communicative practice that supports community collaboration, creativity, and innovative solutions. Dialogue encompasses the communication and information exchange that sustains social capital, but organisational communication scholars emphasise the link between dialogue and a special kind of learning and creativity that can lead to positive change through collaboration (Deetz 1995; Deetz & Simpson 2004; Heath et al. 2006). A constitutive element of dialogue is *having voice*, which according to Deetz (1995) can create new understandings of a given situation that can only become known through the dialogic experience (see also Milam and Heath 2014). Susan Senecah (2004) developed the trinity of voice (TOV) model in which she demonstrated the importance of voice for people's participation in and influence on decision-making processes. As with the processes of communication described in previous sections, voice encompasses information sharing and exchange of ideas that fosters collaboration but also enables individuals to participate in influential decisions in democratic contexts (Milam and Heath 2014). The concept of voice will be revisited in Chap. 5, but at this point concepts such as voice and dialogue offer conceptual tools for unfolding the role of informal and volitional communication processes in development and social change. In this section, I want to draw specifically on the role of communication in social change through the process of community collaboration.

Communication scholars Heath and Frey (2004) describe community collaboration as 'autonomous stakeholders with varying capabilities ... directed toward mutually accountable, typically innovative ends, producing long-term social change at a local level in a cooperative, relatively nonhierarchical relationship that is negotiated in an ongoing communicative and principled process' (p. 194). Collaboration is understood as a discursive and communicative process, foregrounding egalitarian values such as transparency, shared decision-making, and a cooperative ethos (Lewis 2006). It is also understood to encourage innovative solutions, 'that no single party could have envisioned or enacted' (Gray 1989: 16). Dialogue and communication become the shared focus in collaboration. Through reasoning, arguing, and questioning, communication produces shared meanings and understanding about a given situation or problem that can lead to further action (Lewis 2006). The idea of community collaboration has also been explained through the communities of practice (COPs) literature (Wenger 1998). Communities of practice can be described as social learning systems with a focus on 'informal and situated social interaction, rather than on a planned mechanistic process of cognitive transmission … achiev[ing] authentic, motivated learning of what is needed to be known about the complexities of real practice' (Cox 2005: 4 in Mann 2019: 57). They are informal and self-managed, and their members share common interests and interactions through which they form a common identity and understanding (Lave and Wenger 1991). According to Iverson and McPhee (2002), the interaction among COP members takes place through communicative techniques such as the sharing of ideas, stories, knowledge, and insights. Hence, communities of practice form sources of formal and informal, tacit knowledge that is key to social change (Brown and Duguid 2001). Communities of practice reflect a scholarly focus on learning as participation in the social world where people interact with one another, share experiences, and produce new knowledge (Lave and Wenger 1991). A theory and concept that is closely affiliated with the communities of practice refers to the 'networks of practice'. Like communities of practice, networks of practice are bound together by shared practices, identities, and knowledge, but they have weaker ties. As a result, the knowledge that is produced through the members' interactions may also be influenced by the members' exchanges with others outside the network (Hustad and Teigland 2005).

Communities and networks of practice bring to the forefront the importance of social learning in rural contexts and agriculture extension. This type of learning is not based on top-down and linear information diffusion through communication technologies but on iterative reflection that occurs when we share our experiences, ideas, and environments with others (Oreszczyn et al. 2010: 406). The concepts and theories of communities and networks of practice highlight the importance of communication practices that are built on informal, open-ended, and dynamic social relations through which community members negotiate meaning and identity, co-construct knowledge, self-organise, and achieve social change. In a way, these concepts and practices underline those elements that, according to communication for development scholars, deserve more attention in the context of development. Such elements include dialogue and genuine community participation in decision-making regarding community development, and new contexts of collective action and development. Communities of practice have also been approached through the prism of food networks by Alana Mann (2019), who analysed the practices of participation and collective and emancipatory learning among food growers and consumers in relation to food democracy. Studies like this create more fertile ground for unearthing the different possibilities for communication, well-being, development, and social change that can be found in non-institutional contexts such as the global food market.

The insights provided by the theories of interpersonal communication, social capital, and community collaboration reiterate the powerful nature of informal social relations and communication practices and their contribution to different types of development. Through these theories, we are reminded of the multiple benefits—such as coping and a sense of identity and belonging—that individuals can gain through collaboration and of the non-material assets, such as psychological, social, and informational resources. In essence, the tangible and intangible outcomes that are achieved through the social learning and knowledge construction and sharing that occurs in the context of such social relations provide additional analytical purchase to the concept of weak communication for the way we study and talk about the role of communication in sustainable development and social change. Communication practices such as interpersonal communication, social capital, and community collaboration happen in pragmatic, everyday life contexts that differ from the institutionalised programmes that promote development through communication. Yet, the fact that such communication practices can lead to social change is testament that new approaches are needed that embrace surprise and alternative theoretical and empirical investigations of the role that communication can play in sustainable development.

Conclusion

Sustainable development implies a participatory approach to policymaking and implementation (Servaes 2013: 6). To this end, the study of communication for sustainable development and social change demonstrates the progress that has been achieved through communication, with practices and traditions of discourse turning from modernisation perspectives towards community participation approaches and towards greater and more meaningful engagement of communities with communication technologies. Yet, as the critiques presented in this chapter reveal, the field continues to be dominated by institutional efforts that continue to pursue a rather one-dimensional approach to sustainable development that is predominantly dictated by the interests and agendas of the development industry. At the same time, with its critical analysis and the calls for more attention to be paid to the voices of communities, the field has created a fertile ground for the introduction of new lines of thinking and for the discovery of new contexts where communication and sustainable development converge. It is the aim of this book to tap into this fertile ground with the introduction of weak theory and the concept of weak communication. Weak communication serves as a tool that elevates the importance of non-institutionalised communicative practices, and that facilitates the placing and the study of communication in new development contexts. As well as providing a new language, the ideas that underpin weak theory and the concept of weak communication provide a theoretical foundation on which to build an interdisciplinary analysis and deconstruction of sustainable development through communication.

The global food economy and the case of global food networks provide a space that is rich in relational processes and communication practices that lie outside formal institutionalised communication campaigns. It is also inextricably linked with the development of agricultural communities in countries of the Global South. Many of the development initiatives reviewed in this chapter were geared towards creating trade, and opportunities for participation in the global economy for farming communities. The discourses of the global economy have also been instrumental in shaping the institutionalised development efforts that have given rise to concerns among critical scholars. Disrupting these discourses is therefore a key step towards creating new understandings between communication and sustainable development. In the remaining chapters, the sociability and socio-material relations that sustain the global food economy will be explored through different aspects of the food production and consumption process, and through a bottom-up communication perspective. In each case, the aim is to unearth the multidimensional character of sustainable development this time not through prevailing dominant discourses but through the voices of the people.

References

Adler, P. S., & Kwon, S. W. (2002). 'Social capital: Prospects for a new concept.' Academy of management. *The Academy of Management Review, 27*(1), 17–40.

Albrecht, T. L., Adelman, M. B., & Associates. (1987). *Communicating social support*. Newbury Park: Sage.

Anyaegbunam, C., Mefalopulos, P., & Moetsabi, T. (2004) *Participatory rural communication appraisal: Starting with the people*. SADC Centre of Communication for development Harare and food and agriculture organization of the United Nations, 2nd Ed. Rome.

Balit, S. (2012). Communication for development in good and difficult times: The FAO experience. In F. Enghel & K. Wilkins (Eds.), *Communication, media and development: Problems and perspectives* (pp. 105–120). Nordicom Review: Sweden. Special Issue.

Baributsa, D., Lowenberg-DeBoer, J., & Kamarou, D. A. (2010). *Cell phone video for communicating hermetic cowpea storage skills, 5th world cowpea research conference*. Senegal: Dakar.

Bentley, J., Van Mele, P., Zoundji, G., & Guindo, S. (2014). *Social innovations triggered by videos: Evidence from Mali*. Belgium: Agro-Insight Publications.

Bourdieu, P. (1986). The forms of capital. In J. G. Richardson (Ed.), *Handbook of theory and research for the sociology of education* (pp. 241–258). New York: Greenwood.

Brown, J. S., & Duguid, P. (2001). Knowledge and organization: A social-practice perspective. *Organization Science, 12*(2), 198–213.

Burt, R. (1992). *Structural holes: The social structure of competition*. Cambridge: Harvard University Press.

Chambers, R. (1983). *Rural development: Putting the last first*. Harlow: Prentice Hall.

Chambers, R. (1994). The origins and practice of participatory rural appraisal. *World Development, 22*(7), 953–969.

Chambers, R. (1997). Responsible well-being. A personal agenda for development. *World Development, 25*(11), 1743–1754.

Chassagne, N. (2018). Sustaining the 'Good Life': Buen Vivir as an alternative to sustainable development. *Community Development Journal, 54*(3), 482–500. https://doi.org/10.1093/cdj/bsx062.

Chowdhury, A., Odame, H. H., Thompson, S., & Hauser, M. (2015). Enhancing farmers' capacity for botanical pesticide innovation through video-mediated learning in Bangladesh. *International Journal of Agricultural Sustainability, 13*(4), 326–349.

Coldevin, G. (2003). *Participatory communication: A key to rural learning systems, food and agriculture Organization for the United.* Rome: Nations.

Coleman, J. (1990). *The foundations of social theory.* Cambridge: Harvard University Press.

Cox, A. M. (2005). What are communities of practice? A comparative review of four seminal works. *Journal of Information Science, 31*(6), 527–540.

Crowder, L. V. (2000). *National agricultural advisory service.* Uganda: NAADS. FAO/World Bank Formulation Mission Report.

Cunningham, M. R., & Barbee, A. P. (2000). Social support. In C. Hendrick & S. S. Hendrick (Eds.), *Close relationships: A sourcebook* (pp. 273–285). Thousand Oaks: Sage Publications.

Deetz, S. (1995). *Transforming communication, transforming business: Building responsive and responsible workplaces.* Cresskill: Hampton.

Deetz, S., & Simpson, J. (2004). Critical organizational dialogue: Open formation and the demand of "otherness". In R. Anderson, L. A. Baxter, & K. N. Cissna (Eds.), *Dialogue: Theorizing difference in communication studies* (pp. 141–158). Thousand Oaks: Sage.

Dillard, J. P., & Wilson, S. R. (2014). Interpersonal influence. In C. R. Berger (Ed.), *Interpersonal communication* (pp. 155–176). Berlin: De Gruyter Mouton. https://doi.org/10.1109/ICTD.2007.4937388.

Dutta, M. J. (2011). *Communicating social change: Structure, culture, agency.* New York: Routledge.

Dutta, M. J. (2015). Decolonizing communication for social change: A culture-centered approach. *Communication Theory, 25*(2), 123–143. https://doi.org/10.1111/comt.12067.

Easterly, W. (2006). *The white man's burden.* New York: The Penguin Press.

Fafchamps, M., & Minten, B. (1999). Relationships and traders in Madagascar. *Journal of Development Studies, 35*(6), 1–35.

Fafchamps, M., & Minten, B. (2002). Returns to social network capital among traders. *Oxford Economic Papers, 54*(2), 173–206.

Fair, J. E., & Shah, H. (1997). Continuities and discontinuities in communication and development research since 1958. *Journal of International Communication, 4*(2), 3–23.

FAO. (2013). *Information and communication technologies for sustainable agriculture.* Bangkok: Indicators from Asia and the Pacific.

FAO. (2014). *Forum on communication for & community media for family farming.* Rome.

FAO. (2015). *Success stories on information and communication technologies for agriculture and rural development.* Bangkok.
FAO. (2017). *Information and Communication Technology (ICT) in agriculture.* A report to the G20 agricultural deputies, Rome.
Fisher, R. (2013). *Teaching thinking: Philosophical enquiry in the classroom.* London: A&C Black.
Fraser, C., & Restrepo-Estrada, S. (1998). *Communicating for development human change for survival.* London: I. B. Tauris.
Freire, P. (2001). *Pedagogy of the oppressed.* New York: Continuum International Publishing Group Ltd.
Fukuyama, F. (1995). *Trust: The social virtues and the creation of prosperity.* New York: Free Press.
Gandhi, R., Veeraraghavan, R., Toyama, K., & Ramprasad, V. (2009). Digital green: Participatory video and Mediated Instruction for Agricultural Extension. *Information Technologies and International Development, 5*(1), 1–15.
Gergs, H. J. (2003). Economic, social, and symbolic capital. New aspects for the development of a sociological theory of the market. *International Studies of Management and Organization, 33*(2), 22–48.
Granovetter, M. (1973). The strength of weak ties. *The American Journal of Sociology, 78*(6), 1360–1380.
Gray, B. (1989). *Collaborating: Finding common ground for multi-party problems.* San Francisco: Jossey-Bass.
Heath, R. G., & Frey, L. R. (2004). Ideal collaboration: A conceptual framework of community collaboration. In P. Kalbfleisch (Ed.), *Communication Yearbook, 28* (pp. 189–231). Mahwah: Lawrence Erlbaum.
Heath, R. L., Pearce, W. B., Shotter, J., Taylor, J., Kersten, A., Zorn, T., et al. (2006). The process of dialogue: Participation and legitimation. *Management Communication Quarterly, 19*(3), 341–375.
Heeks, R. (2010). Do information and communication technologies (ICTs) contribute to development? *Journal of International Development, 22*(5), 625–640. https://doi.org/10.1002/jid.1716.
Hosman, L., & Fife, E. (2008). Improving the prospects for sustainable ICT projects in the developing world. *International Journal of Media and Cultural Politics, 4*(1), 51–69. https://doi.org/10.1386/macp.4.1.51_1.
Hustad, E., & Teigland, R. (2005). Taking a differentiated view of intra-organizational distributed networks of practice: A case study exploring knowledge activities, diversity and communication media use. In P. van Den Besselaar, G. De Michelis, J. Preece, & C. Simone (Eds.), *Communities and technologies* (pp. 239–261). Netherlands: Springer.
Ikeda, A., & Kawachi, I. (2010). Social networks and health. In A. Steptoe (Ed.), *Handbook of behavioral medicine* (pp. 237–261). New York: Springer.

Iverson, J. O., & McPhee, R. D. (2002). Knowledge management in communities of practice: Being true to the communicative character of knowledge. *Management Communication Quarterly, 16*(2), 259–266.

Keniston, K., & Kumar, D. (2003). *The four digital divides.* New Delhi: Sage Publishers.

Khan, S. R., Rifaqat, Z., & Kazmi, S. (2007). *Harnessing and guiding social capital for rural development.* New York: Palgrave Macmillan.

Kikuchi, M., & Coleman, C.-L. (2012). Explicating and measuring social relationships in social capital research. *Communication Theory, 22*(2), 187–203.

Kleine, D. (2013). *Technologies of choice?* Cambridge: MIT Press.

Lave, J., & Wenger, E. (1991). *Situated learning: Legitimate peripheral participation.* Cambridge: Cambridge University Press.

Law, J., & Urry, J. (2004). Enacting the social. *Economy and Society, 33*(3), 390–410. https://doi.org/10.1080/0308514042000225716.

Lazarsfeld, P. F., Berelson, B., & Gaudet, H. (1944). *The people's choice.* New York: Duell, Sloan & Pearce.

Lee, C., & Sohn, D. (2016). Mapping the social capital research in communication. *Journalism and Mass Communication Quarterly, 93*(4), 728–749.

Lewis, L. (2006). Collaborative interaction: Review of communication scholarship and a research agenda. In C. S. Beck (Ed.), *Communication Yearbook, 30* (pp. 197–247). Mahwah: Lawrence Erlbaum.

Lin, N. (1999). Building a network theory of social capital. *Connect, 22*(1), 28–51.

Lin, N. (2001). Building a network theory of social capital. In Social capital: Theory and research. Lin, N., K. S. Cook, and R. S. Burt (ed.) pp. 3–29. Hawthorne, USA: Aldine de Gruyter.

Lyon, S. (2018). Digital connections. Coffee, agency, and unequal platforms. In T. Schneider, K. Eli, C. Dolan, & S. Ulijaszek (Eds.), *Digital food activism* (pp. 70–88). New York: Routledge.

Makuwira, J. (2018). Power and development in practice: NGOs and the development agenda setting. *Development in Practice, 28*(3), 422–431. https://doi.org/10.1080/09614524.2018.1433816.

Manalo, J. A. IV, Layaoen, M. G., & Parac, E. A. (2009, October 13–14). *Reflections on online communication campaign on rice production: Prospects and inputs for expansion.* Paper presented at the Philippine Extension Network Symposium, Manila.

Manalo, J. A., & van de Fliert, E. (2011). Hypothesizing ICT4D in Philippine agriculture: Deriving from trends, setting directions. *Asian Journal of Agriculture and Development, 8*(2), 17–28.

Mann, A. (2019). *Voice and participation in global food politics.* London\New York: Routledge.

Mansell, R. (1982). The “new dominant paradigm”in communication: Transformation versus adaptation. *Canadian Journal of Communication, 8*(3), 42–60.

Manyozo, L. (2012). *Media, communication and development.* London: Sage.

Manyozo, L. (2017). *Communicating development with communities.* London: Routledge.

McAnany, E. (Ed.). (1980). *Communications in the Rural Thin! World. The role of information in development.* New York: Praeger.

Melkote, S. R. (2003). Theories of development communication. In B. Mody (Ed.), *International and development communication: A 21st-century perspective.* Thousand Oaks: Sage.

Melkote, S. R., & Steeves, H. L. (2015). *Communication for development: Theory and practice for empowerment.* New Delhi: SAGE.

Milam, J., & Heath, R. G. (2014). Participative democracy and voice: Rethinking community collaboration beyond neutral structures. *Journal of Applied Communication Research, 42*(4), 366–386. https://doi.org/10.1080/00909882.2014.911944.

Manalo, J.A. IV, Layaoen, M.G.and Parac, E.A. (2009) ‘*Reflections on Online Communication Campaign on Rice Production: Prospects and Inputs for Expansion*’, Paper presented at the Philippine Extension Network Symposium, Manila, October 13–14.

Morris, N. (2003). The diffusion and participatory models: A comparative analysis. In O. Hemer & T. Tufte (Eds.), *Media and Glocal change. Rethinking communication for development* (pp. 123–144). Gothenburg/Buenos Aires: Nordicom and CLACSO.

Musa, N., Githeko, J., & El-Siddig, K. (2013). Challenges of using information and communication technologies to disseminate agricultural information to farmers in Sudan. *International Journal of Sudan Research, 3*(2), 117–131.

Oreszczyn, S., Lane, A., & Carr, S. (2010). The role of networks of practice and webs of influencers on farmers’ engagement with and learning about agricultural innovations. *Journal of Rural Studies, 26*(4), 404–417. https://doi.org/10.1016/j.jrurstud.2010.03.003.

Portes, A., & Landolt, P. (2000). Social capital: Promise and pitfalls of its role in development. *Journal of Latin American Studies, 32*(2), 529–547. https://doi.org/10.1017/S0022216X00005836.

Putnam, R. D. (1993). The prosperous community: Social capital and public life. *The American Prospect, 13*, 35–42.

Putnam, R. (2001). *Bowling alone: The collapse and revival of American community.* New York: Simon & Schuster.

Quarry, W., & Ramírez, R. (2009). *Communication for another development: Listening before telling.* New York: Zed Books.

Ramalingam, B. (2014). *Navigating 'wicked' problems in development.* Overseas Development Institute. Available at https://www.odi.org/blogs/8801-navigating-wicked-problems-development. Accessed 7 Sept 2016.

Reimer, B., Lyons, T., Ferguson, N., et al. (2008). Social capital as social relations: The contribution of normative structures. *The Sociological Review, 56*(2), 256–274. https://doi.org/10.1111/j.1467-954X.2008.00787.x.

Rivera, M., Knickel, K., de los Rios, I., et al. (2018). Rethinking the connections between agricultural change and rural prosperity: A discussion of insights derived from case studies in seven countries. *Journal of Rural Studies, 59*, 242–251. https://doi.org/10.1016/j.jrurstud.2017.07.006.

Robison, L. J., Schmid, A. A., & Siles, M. E. (2002). Is social capital really capital? *Review of Social Economy, 60*(1), 1–24.

Rodriguez, C. (2011). *Citizens' media against armed conflict: Disrupting violence in Colombia*. Minneapolis: The University of Minnesota Press.

Senecah, S. (2004). The trinity of voice: The role of practical theory in planning and evaluating the effectiveness of environmental participatory processes. In S. P. Depoe, J. W. Delicath, & M. A. Elsenbeer (Eds.), *Communication and public participation in environmental decision making* (pp. 13–33). Albany: State University of New York Press.

Servaes, J. (2013). Introduction: The kaleidoscope of text and context in communication. In J. Servaes (Ed.), *Sustainability, participation and culture in communication: theory and praxis* (pp. 1–24). Intellect: Bristol.

Servaes, J., & Lie, R. (2015). New challenges for communication for sustainable development and social change: A review essay. *Journal of Multicultural Discourses, 10*(1), 124–148. https://doi.org/10.1080/17447143.2014.982655.

Sodano, V., & Verneau, F. (2006, February 8–10). *Social capital and the food system: Some evidences from empirical research.* Paper prepared for presentation at the 99th EAAE seminar 'trust and risk in business networks'. Bonn, Germany.

Solomon, D. H., & Vangelisti. (2010). Establishing and maintaining relationships. In C. R. Berger, M. E. Roloff, & D. R. Roskos-Ewoldsen (Eds.), *The handbook of communication science* (pp. 326–344). Thousand Oaks: SAGE.

Sousa, F., Nicolay, G., & Home, R. (2019). Video on Mobile phones as an effective way to promote sustainable practices by facilitating innovation uptake in Mali. *International Journal of Sustainable Development Research, 5*(1), 1–8. https://doi.org/10.11648/j.ijsdr.20190501.11.

Sulaiman, V., Hall, A., Kalaivani, N., Dorai, K., & Reddy, T. (2012). Necessary, but not sufficient: Critiquing the role of information and communication technology in putting knowledge into use. *The Journal of Agricultural Education and Extension., 18*(4), 331–346.

Tacchi, J., & Tufte, T. (2020). Communicating for change. In J. Tacchi & T. Tufte (Eds.), *Communicating for change: Concepts to think with* (pp. 1–5). Switzerland: Palgrave Macmillan.

Tata, J. S., & McNamara, P. E. (2016). Social factors that influence use of ICT in agricultural extension in southern Africa, agriculture, MDPI. *Open Access Journal, 6*(2), 1–10.

Thomas, P. N. (2011). *Negotiating communication rights: Case studies from India.* New Delhi: Sage Publications.

Thomas, P. N. (2014). Theorizing development, communication and social change. In C. Christians & K. Nodenstreng (Eds.), *Communication theories in a multicultural world* (pp. 213–227). New York: Peter Lang Publishing.

Thomas, P. N. (2015). Communication and social change, making theory count. *Nordicom Review, 36*(Special issue), 71–78.

Touri, M. (2020). The power of weak communication. In J. Tacchi & T. Tufte (Eds.), *Communicating for change: Concepts to think with* (pp. 75–84). Palgrave Macmillan: Switzerland.

Tufte, T. (2017). *Communication and social change: A citizen perspective.* Cambridge: Polity.

Tufte, T., & Mefalopoulos, P. (2009, December). *Participatory communication: A practical guide, World Bank publications.* The World Bank, number 5940.

Unwin, T. (2009). *ICT4D: Information and communication technology for development.* Cambridge: Cambridge University Press.

van de Fliert, E. (2007). Communication and sustainable rural development. In P. N. Thomas (Ed.), *Communication for social change* (pp. 85–94). Brisbane: The University of Queensland.

van de Fliert, E. (2010). Communication and sustainable rural development. In P. N. Thomas & M. Bromley (Eds.), *An introduction to communication and social change* (pp. 89–98). Brisbane: University of Queensland Press.

van de Fliert, E. (2014). Communication, development, and the natural environment. In K. Gwinn Wilkins, T. Tufte, & R. Obregon (Eds.), *The handbook of development communication and social change* (pp. 125–137). Wiley-Blackwell: Chichester.

Van Mele, P., Wanvoeke, J., & Zossou, E. (2010). Enhancing rural learning, linkages, and institutions: The rice videos in Africa. *Development in Practice, 20*(3), 414–421.

van Staveren, I., & Knorringa, P. (2007). Unpacking social Capital in Economic Development: How social relations matter. *Review of Social Economy, 65*(1), 107–135.

Wenger, E. C. (1998). *Communities of practice: Learning, meaning, and identity.* Cambridge: Cambridge University Press.

White, S. A. (1994). The concept of participation: Transforming rhetoric to reality. In S. A. White, K. S. Nair, & J. Ascroft (Eds.), *Participatory communication: Working for change and development* (pp. 15–32). New Delhi: Sage.
Wilkins, K. G. (2000). Accounting for power in development communication. In K. Wilkins (Ed.), *Redeveloping communication for social change: Theory, practice & power* (pp. 197–210). Boulder: Rowman & Littlefield Publishers.
Wilkins, K. G., & Enghel, F. (2013). The privatization of development through global communication industries: Living proof? *Media, Culture and Society, 35*(2), 165–181.
World Bank. (2000). *Panama poverty assessment: Priorities and strategies for poverty reduction.* Washington D.C: World Bank Group Publications.

CHAPTER 4

Communication and Business in the Global Market: Empowering Indian Farmers Through Trade Relations

Introduction

Development discourses reproduce a world we already know, 'a world of production and markets, of good and bad, of developed and underdeveloped, of aid, of investment by multinational corporations, of science and technology, of progress and happiness, of individuality and economics' (Escobar 2001: 162). This chapter argues that hegemonic discourses of development are also integral in the role that the private sector and business is perceived to play in development. However, deconstructing and re-politicising the idea of sustainable development requires recognition of the multiple and non-monetised contributions that business can make. A new language is, therefore, needed to bring to light the less formal and more subtle paths through which business relations can prove empowering. Drawing on weak theory and the concept of weak communication (Chaps. 2 and 3), this chapter introduces the practice of communication in business' role in development. It presents an analysis of the power of interpersonal and dialogic communication between the farmers in South India and SMEs in Europe and North America to generate development opportunities. This is based on original evidence of the social relations and communication practices that connect the two groups.

The chapter begins with a critical analysis of the role of business in development through relevant theories and concepts. Business' role as a development tool and agent has long been recognised by international

M. Touri, *Sustainable Development and Communication in Global Food Networks*, https://doi.org/10.1007/978-3-030-46119-5_4

development agents, leading to a more systematic inclusion of private enterprises in development agendas. Yet, such initiatives are guided by ideas of development primarily as economic growth and by the principles of modernisation theories. The case of Corporate Social Responsibility (CSR) is a representative example of business initiatives that encourage a de-politicisation of development by focusing almost entirely on poverty alleviation through mechanistic practices. This idea is expounded further through the Bottom Billion Capitalism (BBC) and the Bottom of the Pyramid (BOP) cases. These are initiatives that were designed to support economic growth of the poorer communities in the Global South through the creation of business opportunities for income growth and the consumption of products that were specifically designed for the poor. However, once again these efforts demonstrate limited understanding of well-being and development, framing it as a condition that can be achieved simply through solutions offered by products and services. A more inclusive and participatory conceptualisation of the business-development nexus can be found in the case of social entrepreneurship (SE). Although not exempt from criticism for advancing a free market ideology and a modernisation approach to development, SE involves relations and communication practices that can pave the way for new discourses around the role of business in development.

The business-development cases described in the first section of the chapter are typical of a discourse of development and empowerment as income generation and poverty alleviation. This is a discourse that not only encourages an incomplete vision of the role of business in development, it also conceals the complexity of the economy and the multidimensional character of well-being. The second part of the chapter presents an alternative and more malleable conception of business relations and the different paths through which they can contribute to development. This is built on theoretical propositions about the inadequacy of real incomes as an analytical metric for assessing development and the need for a new language and a new way of thinking in the study of development. A new way of thinking is further explored through the non-monetary role that the five private enterprises from Europe and North America play in the development of the farming communities in Organic Wayanad.

Market-Based Solutions and the Role of Business in Development

The role of business, and of the private sector more broadly, has received significant attention during the last three decades as a result of neoliberal globalisation, technological progress, and the growing transnationalisation of production and finance (Reed and Reed, 2009). According to Blowfield (2010) and Blowfield and Dolan (2014), this role can be approached either as a 'tool' or as an 'agent'. As a development tool, commercial enterprises contribute to the public good through their engagement in commerce and trade and through wealth and value creation for their owners. However, this growth-driven development is not an intentional development initiative, since it results from the type of activities that any business is expected to be involved in. Development takes place through an unintentional process that refers to the usual growth of national economies, and the improvement in welfare and human well-being that this creates. Commercial enterprise plays a key role through various routes such as investment by foreign companies, the growth of local entrepreneurship, or trade relations among countries.

Instead, when organisations engage in activities that are designed to create a calculated development benefit beyond their own profit, they assume the role of the development agent. Here the focus is on the intentional commercial activities undertaken to promote global development goals and support the economic prosperity of poorer countries. Businesses can also join forces and combine resources with development agencies and civil society organisations to develop programmes for health and education, environment protection, and human rights. This intentional development role was the outcome of the increasing criticism and scrutiny that several companies came under during the 1980s and following their involvement in human right abuses and environmentally damaging activities. Since then, businesses began to take a more proactive role, signing up to voluntary codes of practice and looking for development opportunities particularly in countries of the Global South where investments were made to support marginalised communities. As a development agent, business is expected to 'consciously and accountably strive to address poverty and marginalisation' (Blowfield and Dolan 2014: 26), becoming accountable not only for financial performance but also for development activities and outcomes. As companies began to play a more instrumental role in development, international organisations including the United Nations

Conference on Trade and Development (UNCTAD), the United Nations Development Programme (UNDP), and the World Trade Organization (WTO) introduced agendas on the role of business in development, with the focus being particularly on alleviating poverty (Kolk and van Tulder 2006). The discourse of social sustainable development and corporate social responsibility has now become universal as businesses are increasingly being encouraged to participate in private regulatory initiatives for social and environmental responsibility, and to voluntarily adopt practices that will improve their social and environmental performance (Blowfield 2010).

For the most part, the contribution of business in development has been theorised in line with dominant ideas of development as economic growth and modernity. In a similar vein to the theories of modernisation in development, where traditional societies become 'developed' by adopting the institutions and practices of modern economies, the role of businesses is largely understood through the creation of opportunities for economic growth and inclusion of the 'poor' in global economy. In other words, and to follow Easterly's (2006) division of the development industry into 'planners' and 'searchers', the role of business has been largely explained in line with the definition of 'planners', who set big goals and adopt a top-down 'transformational' approach to development, but neglect the complexity of ground reality; instead, searchers tend to be more process-oriented, paying attention and listening to the people on the ground.

The increasingly more active involvement of corporations in development has also been driven by advocates of neoliberal globalisation to counter the growing resistance to economic liberalisation (Reed and Reed 2009). Interestingly, commercial activities are believed to offer an alternative answer to the failure of modernisation-theory-driven development institutions to support sustainable development in countries of the Global South (Roxas and Ungson 2011). However, in reality, market-based solutions to development are still more likely to follow the logic of modernisation theory, approaching development as an economic process that focuses on giving 'access' to goods and services while neglecting the need to build knowledge and capabilities of marginalised communities.

Business, Poverty, and Development Economics in CSR

Assessing the contribution of businesses in development inevitably brings to the fore questions around understanding and defining development. For the most part, business is increasingly being promoted as a solution to poverty, which constitutes the primary condition dictating the role of business as a development agent (Blowfield 2010). When it poses risks to a company or industry, when it offers a favourable return on investment, and when it causes inefficiency and corruption, poverty becomes a driving force in business' agent role. Yet, market-driven solutions to development that are designed with the aim to alleviate poverty—by providing employment and affordable services to poor communities or offering investment solutions—reflect a reductionist and inadequate focus on investment and financial aid to pull communities out of poverty. Besides, in most cases where poverty is the focus, contributions to development derive mostly from standard business-as-usual activities rather than from development-oriented philanthropic projects. Based on a review of existing research by Newell and Frynas (2007), businesses help alleviate poverty in countries of the Global South through increasing employment and access to goods and services among the poor, as well as through private investments and tax contributions. Although such practices can generate development gains not just through investment but also through the introduction of cleaner production processes and advanced technologies, they have less to do with the social standards, responsibility, and good intentions of corporations and more with their focus on profit and economic growth.

Business' focus on poverty alleviation as a form of development is at the core of the Corporate Social Responsibility (CSR) business model, a self-regulating model that helps companies be socially accountable and simultaneously contribute to international development. CSR emerged as a PR tool to deflect criticism and be seen to be doing good. Although the initial focus of CSR was on corporate conduct that impinges on social, environmental, and human rights issues, its scope has now broadened to include the role of business in relation to poverty reduction and tackling social exclusion in the developing world (Newell and Frynas 2007). As a development tool, CSR is predicated on the idea of the market playing a critical role in alleviating poverty. This is a distinct role to CSR as a business tool and has been embraced by international organisations and national development agencies with the expectation that businesses can make a significant contribution to international development (Newell and Frynas 2007).

The strategies of CSR initiatives for addressing underdevelopment and poverty encapsulate a rather mechanistic approach to development. Development is framed as the outcome of practices including innovation and investment, while economic, social, and political problems are reduced to technical compartments that are treated separately from one another, leading to the depoliticisation of development as a whole (Escobar 2001). By approaching the benefits of CSR initiatives for poorer and marginalised groups through corporations' day-to day practices and output, most readings of CSR in development undermine the importance of different forms of exclusion that keep marginalised communities in poverty. Gender issues are a point in case here, since the needs of women are seldom considered in the design, enforcement, and evaluation of CSR codes of conduct and regimes of accountability (see Smith and Dolan 2006; Prieto and Quinteros 2004). More importantly, as Pearson (2007) explains, it is the deeper political and structural roots of gender inequality that most CSR initiatives ignore. In this way, codes of conduct are set up to represent a universal and neutral solution to problematic labour conditions while the social and political causes of such conditions are removed from consideration (Blowfield and Dolan 2008). This depoliticisation is the ultimate devaluation of the non-monetised role of private enterprises. This is further exemplified in the case of Bottom Billion Capitalism and the notion that there is profit to be made by targeting products and services at the 'bottom of the pyramid' (BOP) (Newell and Frynas 2007).

Bottom Billion Capitalism and the Bottom of the Pyramid Case

'Bottom Billion Capitalism' is one of the most talked-about development initiatives of the private sector in the twenty-first century. The term refers to an array of business and organisational strategies and techniques through which a more 'creative' type of capitalism emerges that aims at generating and sustaining social change through finding solutions to social problems. Although CSR features as one of the prominent solutions of this creative capitalism, the focus of attention has been primarily on new models of partnerships, consumption, and entrepreneurship that aim to bring the world's poor into global markets, recasting poverty as an opportunity. Blowfield and Dolan (2014) refer to a new architecture of global development altogether, which Roy (2010) describes as 'millennial development', while Banks and Hulme (2014) refer to a new generation of philanthrocapitalists that aim to transform philanthropy into a lucrative industry. The term 'bottom billion' has been popularised by two scholars in particular, Paul Collier and C. K. Prahalad, both of whom advocate a new approach to foreign aid to support the four billion poor at the so-called base of the pyramid. While Collier (2007) champions for more funding, trade, and military assistance, Prahalad (2009) makes the case for a market-driven model that comes in the package of entrepreneurship and consumption amongst populations living in extreme poverty.

The term 'Bottom Billion' is also associated with the term 'Bottom of the Pyramid' or 'Base of the Pyramid' referring to the poorest socio-economic group in the economic chain. This typically includes the four billion people living in developing and emerging economies with a per-capita income of under \$2000 per year (Hall et al. 2012; Hammond et al. 2007). Management scholar C. K. Pahalad developed a vision for an innovative type of capitalism where market-driven strategies would help relieve poverty and facilitate social transformation in less advanced countries. Proponents of BOP initiatives focus on the importance of moving away from portrayals of the poor as passive recipients of and dependent on external foreign aid. Instead, on the one hand, these communities are recast as 'agents' and 'entrepreneurs' and responsible for their own development through investing their assets, skills, and knowledge in market-building opportunities. In this case, business is recast as development; and poverty is overcome through the enterprise and entrepreneurship of the poor. On the other, through their collective consumption power, they transform into a business opportunity for corporations (Blowfield and Dolan 2014; Prahalad 2009). Specifically, as the BOP market consists of billions of potential customers, it presents a unique opportunity for businesses to introduce these customers to commercial markets and convert BOP into a profitable endeavour. Despite its weak purchasing power, the BOP market still represents a large segment of the world's population; and finding profitable ways to meet the needs of these groups can attract far more capital than through aid. It is particularly the low-cost products and services that can provide a win-win solution as they can increase the well-being of the poor and the profits of the private sector (Prahalad 2009; UNDP 2008). BOP models also seek to challenge dominant marketing assumptions about the purchasing power of the poor and their willingness to pay for high-quality services and products (Kuriyan and Ray 2008). As a result, investment in bottom-of-the-pyramid businesses has soared over the years, with microfinance being the most popular so far (Economist 2009). Microfinance or microcredit involves the provision of loans to poor borrowers who typically lack security. Other initiatives involve inclusive business, fair trade, and systems that bring new products and services to consumers in the less developed parts of the world.

Unsurprisingly, the BOP models have been seen with growing doubt and suspicion by scholars that point to Prahalad's unconvincing argument about how exactly such ventures could ameliorate poverty, especially by providing greater choice of products to the poor; or how the poor could take advantage of enterprise-based opportunities, especially when they are treated as consumers rather than as producers (Karnani 2007). Seeing the poor as a homogenous group means that there is indeed little understanding of whether and how they could really benefit from business opportunities on new products, while developmental benefits of BOP remain crudely assessed (Banerjee and Duflo 2007; Blowfield 2012). For instance, in their case study of the Akshaya project, Kuriyan and Ray (2008) examined the implementation of BOP principles in Kerala, India, focusing on the use of ICT kiosks by the local population. Their findings showed that the primary customers of ICT kiosks are not the poor but those who can afford to pay for relevant applications on an ongoing basis. Others point to the fact that 'capitalist' solutions that are underpinned by neoliberal valorisation of private enterprise cannot be seen as a genuine development alternative (Banks and Hulme 2014).

A key challenge posed by BBC and BOP initiatives lies in their overly economic determinist measurement of development. First of all, by basing the development of the poor communities on income growth and economic returns, these initiatives promote a very narrow understanding and evaluation of well-being as a condition that can be achieved through solutions offered by particular products and services. This is aptly manifested in models such as fair trade where the solutions envisaged by the companies and consumers involved do not necessarily match the needs of the farmers' communities. Research on Ghanaian cocoa producers reveals how the allocation of fair trade's social premium on health and education discounted the preferences of producers for better roads, places of worship, or cash transfer (Blowfield and Dolan 2014). Not only are alternative ways of measuring development and well-being entirely ignored so are the voices of the very people that these solutions aim to help.

BBC initiatives are driven by economic-oriented development solutions that need to correspond to commercial imperatives and Western perceptions of poverty and underdevelopment. Examples of such solutions include Internet access and availability of goods distributed through BOP networks (soap, skin whitening cream, etc.). On the one hand, such solutions would normally fall into place after basic needs were fulfilled; on the other, they frame poor communities as consumers whose needs are defined and fulfilled by the market (Bellemare 2014; Blowfield and Dolan 2014). Ultimately, the whole notion of well-being and community capability building is reduced to income generation, while community skills remain unexplored and community needs unfulfilled. The balance between the market segment and social cause element, which BOP initiatives fail to achieve, has become the mantra of social entrepreneurship, a theoretical and practical avenue that seeks to prioritise community needs and the development of community skills and capabilities.

Development Through Social Entrepreneurship and Microcredit

Social entrepreneurship can be defined as 'a process by which citizens build or transform institutions to advance solutions to social problems, such as poverty, illness, illiteracy, environmental destruction, human rights abuses and corruption, in order to make life better for many' (Bornstein and Davis 2010: 1). Nicholls (2006) suggests putting 'boundaries' around the definition of SE, and he defines it by its two main elements, the strategic focus on social impact and the innovative approach to achieving its mission (2006: 13). Along similar lines, Rowena Young emphasises the contribution of SE in solving local problems that are not addressed by other means (Nicholls 2006).

The actors most commonly involved in such endeavours are companies/businesses and entrepreneurs from the private sector as well as educational institutions and members of the community. Social entrepreneurship is believed to encourage cooperation between those different actors, placing the emphasis on the community needs. At the same time, businesses not only incorporate a social, cultural, and environmental responsibility in their economic activity but also measure their performance based on their positive impact on society. In this case, development is supposed to emerge from the creative problem solving of community groups that is facilitated by the partnership between the state, the public, and the private sector (Nega and Schneider 2013). To qualify as a development practice, social entrepreneurship needs to fulfil specific criteria, including creativity, the creation of a new and original idea for solving a social problem, entrepreneurial value, and social impact (McAnany 2012). Although, like BOP activities, social entrepreneurship focuses on the provision of products and services to poorer communities, the difference lies in the latter's aims to fill the vacuum left by the economic and social institutions that fail to satisfy human needs. The focus is on inventing new systems and solutions that first and foremost aim at creating social value creation, while profit and economic value creation become secondary and mostly for purposes of sustainability and self-sufficiency. This is a logical consequence of the fact that the 'customers' that these endeavours cater for are normally poorer communities that often lack the financial means to pay for services and products (Seelos and Mair 2017).

The practice of social entrepreneurship is not a novel approach, yet, the term and concept became widely spread more recently during the 1980s and 1990s and thanks to two early proponents, Bill Drayton and Nobel Peace Laureate Muhammad Yunus. Drayton is the founder of Ashoka, an NGO that sponsored the ideas of talented entrepreneurs for social innovation. Yunus, an economist in Bangladesh, is credited with institutionalising the concept of microfinance, following the founding of the Grameen Bank, an institution that provided small loans to impoverished people or groups of borrowers—called microentrepreneurs— who were traditionally unable to get financial help from mainstream institutions. These small loans allowed them to pursue entrepreneurial ventures, rendering Yunus' innovative idea a global phenomenon, especially following its positive effect on women, a vulnerable group that was also restricted by cultural norms and barriers (McAnany 2012; Roxas and Ungson 2011). Such was the alleged success of the Grameen Bank in mitigating poverty that in 2005 the United Nations declared the International Year of Microcredit. Microfinance is promoted also by USAID and the World Bank and supported by foundations and individuals, including Google.org, the Bill and Melinda Gates Foundation, and eBay founder Pierre Omidyar (Bornstein and Davis 2010; Bruck 2006).

Compared to the cases of CRS and BOP, SE paves the way for a less top-down understanding of the role of business and development, and one that incorporates the role of communication. McAnany (2012) also draws attention to the more participatory approach in its conceptualisation, mainly because of its local focus in solving problems, as only after an idea has been tested and works locally, does it spread to the whole country or particular arena of application. Hence, although the term in itself may reveal some contradictions, with entrepreneurs' focus on efficiency and profit-making being seen as incompatible with social change, social entrepreneurship actually involves relations and communication practices that can pave the way for new discourses around the role of business in development. Even if recent theories of SE do not stipulate communication as an essential part of the SE model, in practice, both the cases of Yunus and Drayton incorporated interpersonal and technology-based communication in their structures. Interpersonal communication might be the most basic form of communication and often taken for granted but is central to SE projects (ibid). Nevertheless, SE initiatives have not escaped criticism, with regard to balancing financial success with non-financial incentives, especially as banks and financing organisations are manipulating the poor by imposing higher interest rates to reflect higher transactional costs (Roxas and Ungson 2011). Another question concerns how to measure the social impact of social entrepreneurship. In fact, McAnany (2012) explains that one of the reasons for the recognition of SE as part of the development agenda was the emphasis on impact that was inherited from business entrepreneurship. However, although business impact is easily measured through profit, the same does not apply to social impact, a problem that remains unresolved.

As with the case of other business initiatives, social entrepreneurship and microfinance have been criticised for promoting an understanding of development as the advancement of a free market ideology, which resonates perfectly with modernisation theories of development. Brigg (2006: 76–77) is critical towards Yunus' approach to microcredit as a development tool for being entirely driven by a neoliberal and economistic ideology, for elevating the entrepreneurial subjectivity above other subjective modalities and promoting individual initiative and determination as the solution to poverty and underdevelopment. Proponents of this approach generally fail to 'reach the poor, understand their environments, and to recognize their potential to be active participants in the economy' (Fernando 2006: 7). Moreover, by focusing on provision of capital to improve people's situations and increase economic growth, microfinance essentially represents a micro version of the dominant economistic development approach and is congruent with the Western worldview and framework of global capitalism (Brigg 2001; Vakulabharanam and Motiram 2007).

The problems with microfinance as a route to development are both practical and ethical. Microfinance works on the basis of the classical economic position that it offers a win-win solution to creditors and debtors, enabling the latter to invest in capital that will boost growth and allow them to repay the loans. Schanbacher (2010) highlights at least two issues here. On the one hand, in many cases, poor people end up using the funds for emergency purposes rather for investing in entrepreneurship. On the other, microfinance imposes relations that are based on competition, efficiency, and profitability and inserts a deeply capitalist order in ideas and processes of development, discounting the fact that, especially in the case of food production, farmers are driven not just by economic benefits but also by environmental and cultural motives. Schanbacher (2010) refers to the case of the partnership between the Grameen Bank of Bangladesh and the U.S.-based Monsanto Company, where microfinance funds that were made available to farmers were eventually spent on Monsanto's products benefiting Monsanto more than the farmers' entrepreneurial capacity or farmers' knowledge (pp. 27–28). Cases like this essentially reflect a broader critique towards a certain definition of poverty and underdevelopment that is promoted by institutional development and poverty alleviation programmes that claim to be based on an assessment of the poor people's needs. But, in reality, the needs that these programmes seek to address reflect a certain conception of poverty that can be eradicated through the production of economic goods and services but which fails to consider how the poor perceive their own needs (Rahnema 1991).

Re-visiting the Poverty-Development Nexus and the Role of Business in Development

The cases of social entrepreneurship and microcredit are typical of a discourse of development and empowerment as income generation and poverty alleviation that encourages an incomplete and managerial vision of the role of business in development (Blowfield 2010). It is only when our view of the poverty-development nexus is expanded that a different understanding of the role of business in development can also emerge. Efforts to re-define the idea and state of poverty cannot but acknowledge the seminal work of Amartya Sen. In his most influential publication

Development as Freedom, Sen (1999) raises a powerful argument about the need to shift our attention from low income to deprivation of capabilities as a more important criterion of disadvantage than the lowness of income. In the context of the market, the underlying prerequisite for capability maximisation is freedom to participate in economic transactions. When people are denied the opportunities of participation in the market mechanism, they are faced with the type of 'unfreedom' that affects their well-being.

The significance of Sen's argument lies in the distinction that he makes between the type of development that relates to freedom of participation in the market, and the development that results from the market's capacity to expand income and wealth. In other words, he draws a distinction between the value of freedom and the value of utility associated with economic opportunities. In this case, by focusing on income generation, BOP initiatives emphasise what Sen (1999: 27) describes as 'culmination' outcomes, ignoring the importance of the *process* of getting there, and leading to an underappreciation of the full role of the market mechanism. Fernando (2006: 11–12) is critical of Sen's conceptualisation of development for de-politicising poverty and providing numerous possibilities for neoliberal institutions to dictate the discourse of empowerment through theories of capitalist economy rather than social change. However, one of the strengths in Sen's approach is its emphasis on the importance of people's freedom to decide where to work and what to produce and consume, which encapsulates the merit that derives from the full process of the market and not just from the outcomes.

Sen's position about the inadequacy of real incomes as analytical metric for assessing development and satisfaction of needs has already informed attempts to reconceptualise poverty and economic development in the context of BOP and social entrepreneurship. Ansari et al. (2012) and McAnany (2012) have argued for an alternative framework to evaluate BOP interventions. Based on the broader conception of poverty as the lack of capabilities that people have reason to value (Sen 1999), McAnany (2012) draws attention to the complex relationship between income and development, where income should not be approached as an end in itself but as an instrument for achieving other goals. Along similar lines, Ansari et al. (2012) combine insights from the Capability Approach and social capital to argue that BOP ventures can lead to real development if they can increase social capital leading to capability building. Drawing on anthropological assessments of the role of social capital in poverty alleviation,

Ansari et al. (2012) identify network ties, trusted relationships, and shared beliefs and norms as amongst some of the key components of social capital that can lead to capability development. In fact, closeness and trust have been recognised as integral to the relational social capital that sustains economic transactions and contributes to community development (Bolino et al. 2002). These components also find expression in Sen's identification of the powerful systems of values and norms and the non-profit motives that capitalism is dependent on. For Sen (1999: 262–264), behavioural ethics and trusted relations are a crucial ingredient for market success and are integral to the process of the market mechanism that leads to development. Along similar lines, Ansari et al. (2012) suggest that the trusted relations and values that underpin social capital can be the route through which BOP ventures can eventually lead to capability building. This is based on the fact that social capital can affect the necessary conditions for capability development; and such conditions include willingness to share knowledge and dissemination of skills.

Social entrepreneurship models can also embed a relational social capital with personal relations and technology-based communication playing an integral role in the productivity and social change strategies of such models. McAnany (2012) makes a rather compelling observation about technology-based communication being a key component in the social change strategies of Drayton and Yunus, either through the use of telecommunication (cell phones) or web-based social marketing. Moreover, he draws attention to the integral role of interpersonal communication in activities such as teaching, training, and marketing that are central to SE. However, except for social capital and capability building, the different development opportunities that these communication practices may be creating in business relations has somehow escaped academic attention. There are many reasons for this omission, but there is one explanation that I want to put forward: the role of language and discourse. The language of business promotes a certain kind of development; it articulates poor people's empowerment within the boundaries of a hegemonic conception of capitalism as a uniform and totalising process, and obstructs the creation of new foci around which new forms of knowledge can emerge (Blowfield 2010; Escobar 2001; Gibson-Graham 2006). As a result, these discursive constructions of business and development conceal processes and relations that sustain economic transactions and can bring to light the multiple and diverse ways in which business can become the route to development and empowerment, especially for marginalised communities.

The dominance of econoncentric discourses in the academic study and understanding of business' role in development is another symptom of a modernisation ideology promoted through knowledge generation institutions, and the creation of certain forms of knowledge around issues of development and poverty. The way development is envisioned by CSR and BOP proponents overlaps with the corporatisation and privatisation of the development sector, as agencies and donors adopt the practices and language of business, prioritising growth, targets, and returns while ignoring the need for communities to build local ownership and sustainability (Enghel 2015). In this way, for both development agencies and businesses, development is approached based on the idealistic needs and goals of richer countries, which are normally achieved as a by-product of higher incomes but are nowhere on the list of priorities of the less developed and poorer parts of the world (Bellemare 2014).

Yet, the economic discourse permeating the role of business in development is also a reminder of the performativity of knowledge and the power of social science to enact the world that it studies. It is possible that the language of business can co-exist and share the space with alternative languages that reveal the different shapes of the economy and new possibilities and discourses for the role of business in development. One such alternative language is the language of weak theory and weak communication. Extending the theory of weak communication to the role of business in development marks an attempt to change the order of discourse and the knowledge that we produce. In the following sections, the power of weak communication is exemplified through the business relations that underpin global food networks and the relations between the farmers of Organic Wayanad and five of their buyers from Europe and North America.

The Role of Business and Trade Relationships in Global Food Networks

The role of business in the organic and fair trade movements has been the focus of debates, especially in relation to their detrimental effects of the processes of commoditisation and mainstreaming on producers. While, in their initial stages, the two movements comprised small-scale retailers and businesses, the establishment of certification systems and growth in demand has gradually led to the expansion of the distribution networks for both organic and fair trade products, allowing the entrance of big corporations, distributors, and retail outlets. The range of participants in organic and fair trade networks has therefore become much more diverse, including NGOs, larger corporations and retailers, and small-medium enterprises (SMEs). Although all private enterprises, independent of size, are essentially profit-seeking entities, it is important to distinguish between giant corporations and mainstream, profit-oriented companies that tend to conform to minimum organic and fair trade standards, and other—mostly smaller—ethically minded and ideological enterprises as well as movement-oriented importers. For instance, the majority of fair trade organisations in the UK are SMEs with a socially and environmentally responsible philosophy, and although they form networks based on a traditional business model, they play a key role in delivering social benefits to communities (Davies and Crane 2010; Davies and Ryals 2010). More importantly, the heterogeneity and diversity of global food networks in terms of strategies, actors, and locations suggests that the private enterprises involved should also be approached as a heterogeneous group of actors that engage in a multiplicity of economic relations.

Here, the study of Sarah Lyon (2006) on the role of coffee roasters and importers in a cooperative in Guatemala is a case in point, exemplifying the crucial role of these actors in sustaining the fair trade movement and supporting food growers' development. Coffee roasters' willingness to make regular trips in search for high-quality speciality coffee enables them to forge close and long-term relationships, observe and understand the work of producers, and strengthen their trade relationships. More crucially, the close and steady contact between roasters and cooperatives enables the latter to improve their communication skills and their business practices, while they also educate their members regarding foreign market preferences (see also Gereffi 1999; Ronchi 2002). In a different study, Hassler and Franz (2012) and Franz and Hassler (2010) explored the business relationship between a South Indian tribal community that produced organic pepper and a commercial company that sold the product in Germany. In this case, the buyer depicted an ecological responsibility and created a space for the tribal producers' active participation in the construction of knowledge and narratives through which they and their products were represented to the foreign consumer, creating additional value for these products. The authors emphasise how, although such behaviour is normally linked to fair trade organisations, the company's willingness to refrain from exercising power is proof of the need to re-work the boundaries between the conventional and the alternative and reconsider the social contribution of private enterprises in AFNs.

The role of private enterprises in empowering poorer communities in more ways than just income generation is indicative of the diversity of economic exchanges that exist across the organic and fair trade commodity chain. The examples mentioned above are also indicative of the more social and less formal relations that sustain economic activity and can contribute to the development of farming communities. But in order for these development opportunities to become visible and understood, a new way of thinking that moves beyond 'strong theory' dominant approaches is needed. In this chapter, the field of communication for development and social change provides theoretical tools that enable a new, 'weak theory' approach to business' contribution to development and social change through communication. Here, I approach trade relations through the concept of weak communication to unearth the development opportunities that become possible in the global food market when attention is shifted to interpersonal communication, intangible benefits, and localised contexts. The relations that the farmers of Organic Wayanad develop with their foreign buyers provide such a context and are a testament of the subtle but powerful effect of non-mediated and interpersonal communication that sustains these partnerships. This is demonstrated through the different types of empowerment and development that become possible through these relations.

Empowerment and Market Development Through Face-to-Face Communication

Previous research has demonstrated the role of exporting companies and supporting organisations in helping organic farmers increase their knowledge and skills. For instance, in a study of organic farmers in Tanzania, Bakewell-Stone et al. (2008) reported on the dramatic improvement that the monthly visits of field staff from the exporting company had on the productivity of the farmers. What makes such practices significant is the capacity and skills that farmers can learn through the training offered during such visits. Yet, what remains less visible is the crucial role played by the interaction that is embedded in the actual buying and selling process and which can contribute to farmers economic growth but also their human development and well-being more broadly. This was the experience of the farmers in Wayanad thanks to the frequent visits made by their European and North American buyers. In fact, all five buyers interviewed for this research work closely with the farmers' group. They provide training such as for pulping and drying coffee beans and information about consumer demands. Through the buyers' visits, farmers are engaged in a continuous development process as they build new skills and knowledge that can gradually lead to new export opportunities. This learning and development process was recognised by the farmers, as the following quotes demonstrate:

> Some of them make a point of visiting us at least once a year; and every time, I learn something new, whether it is about new ideas for developing products or trends that are working well abroad. These ideas are like a seed that is planted in my head (Farmer A, Group Interview 5).

> There are certain products of high quality that grow here in Wayanad, like turmeric; but not many farmers were cultivating it. When buyers started asking [for it], we understood there was consumer demand, and it was worth investing in it. We then started selling to local consumers too (Farmer B, Group Interview 6). (see also Touri 2018)

More importantly, capacity building and empowerment can happen in an even more nuanced manner through the horizontal interaction and face-to-face communication that takes place between farmers and actors, such as exporters, in the course of their trade relations. In the case of the Wayanadan farmers, this interaction benefited both sides thanks to the continuous communication and information exchange that took place during the buyers' visits. Both the farmers and the buyers stated that they had to spend a considerable amount of time during each visit experimenting with different processes and machinery to achieve the required level of product quality. This experimentation and trial-and-error approach was a learning process that yielded economic benefits for both.

Identifying observable connections between face-to-face communication and development in the form of capacity building and economic returns, as in this case, can occur through the reading of these relations as more than just 'training' that is offered from the expert to the farmer. This type of reading is in line with the principles of weak communication, which offers descriptive sensitivity and can explain one specific element, such as the process through which the buyer-farmer interaction can prove empowering. Behind this interaction lie more complex relationships of exchange that nurture a deeper understanding of each other's work, which can also encourage more trusted relations. Two of the European buyers explained how meeting the farmers frequently in person not only improves their communication and helps avoid procedural misunderstandings; it also enables them to better understand and appreciate each other's needs and requirements, which is crucial for building long-term partnerships. This interpersonal communication between the two actors becomes an indispensable component of their market exchanges since it is through their effort to develop understanding and trust that they can build a mutually beneficial collaboration in economic terms.

A note also needs to be made about the importance of 'size', and how the small size of these buyers matters. On the one hand, associating small enterprises with ethical behaviours would be highly questionable (Samers 2005). On the other, when it comes to maintaining closer relations, direct interactions, and frequent communication with suppliers, size cannot be ignored. It determines the buyers' flexibility and adaptability, the size of their teams, and their relationships with suppliers as well as customers. As the chairman of the farmers' group also explained, small companies that work without intermediaries are willing to devote time to talk to the farmers and to their customers directly, helping build a valuable bridge between the Indian producer and the foreign consumer. The same perception was found even among the buyers, as the words of this buyer show: 'These farmers are actively trying to build a relationship which I appreciate. This relationship building can be a bit more long-term when a company [like mine] is small, as the personnel do not change that rapidly' (Buyer A) (see also Touri 2016, 2018). Effectively, the buyers' size becomes a decisive factor in their communication and the knowledge that farmers can draw from them about foreign markets.

The importance of size can also be seen in the farmers' experience with one of their bigger buyers. Although the company offered training, they adopted a more top-down and less horizontal communication approach, which along with the lower level of engagement that the company showed with regard to the farmers' day-to-day challenges led to communication gaps, misunderstandings, and dissatisfaction. The company claimed they never really received the product quantity they had asked for. They also admitted they were aware that the farmers had to deal with weather issues that had caused damage to their crops, but this knowledge did not stop them from turning to a different supplier group in India, which they described as 'better businessmen' (see also Touri 2016). To a certain extent, this experience reinforces the significance of the interpersonal communication—which was less prominent in this interaction—and its correlation with the size of a company.

Marketing Product Biographies

The marketing and promotion of product biographies is another area where farmers have been able to progress thanks to their interaction and close relationship with certain buyers. Marketing has historically been one of the key areas where improvement would enhance farmers' productivity and income, along with training, agricultural extension, and credit, all of which rely on communication and the availability of information. But as McAnany (1980) states, marketing depends on information in a more intimate way. Communication technologies, such as mobile phones, and more traditional media, such as radio, have provided opportunities for up-to-date information on consumer needs and marketing techniques, as well as for the actual marketing and promotion of agricultural products (Mugwisi 2015; Nakasone et al. 2014). The interviews with the farmers in Wayanad and their buyers demonstrated how close relationships can also provide opportunities for information sharing on marketing-related knowledge, while buyers can function as marketing channels of the farmers' products. In fact, the findings from the interviews reveal how the interaction between the two parties can produce information, knowledge, action, and development that is normally associated with development communication programmes implemented in the field of agriculture. A telling example is the way in which the long-term and regular communication between the two parties has contributed to the farmers' identification of a specific need, the need to understand and promote the provenance and biography of unique products, such as the Wayanadan TGSEB or Tellicherry Garbled Special Extra Bold pepper, a product that has drawn traders from across the globe, including Vasco Da Gama, to Wayanad. The pepper is known for its flavour and taste which, according to the farmers, is owned to the unique soil and climate of Wayanad that gives a special flavour and taste to the local products. Several farmers also commented on the role that the greenery, medicinal plants, and mountains in Kerala play in the composition of the soil and the flavour of certain products. Food provenance and geographical knowledge has special importance in the

Image 4.1 A pepper variety that grows in Wayanad

marketing of food products, for which consumers are willing to pay a premium price (Image 4.1).

However, several of the farmers who were interviewed admitted they were unaware of the marketing value of these qualities. One explanation can be found in the farmers' cultural embeddedness and their shared understanding of the economic value of their products and the consumer perceptions of them. As one farmer stated: 'I literally live with my farm, I look after my plants like my children and I let them grow as naturally as possible; this is my whole life ... but I never thought that would be interesting for any consumer to know' (Farmer C, Group Interview 6). More importantly, those who were aware of the economic value of provenance had gained this knowledge and understanding through their interactions with the buyers. One farmer explained that before they started working with certain buyers, no one had raised these issues before '...we are so involved in these practices on daily basis, we don't have the time to think in this way' (Farmer D, Group Interview 6) (see also Touri 2016). Hence, the close relations and regular communication with buyers paved the way for identification of a specific need and led to action in the form of better negotiation strategies and acquisition of new buyers. Some of the farmers explained that thanks to their interactions with the buyers, they had begun to think and talk more about their unique farming methods, which led to new business partnerships with exporters.

The interviews with the buyers were also revealing in terms of the capacity of businesses to engage into horizontal and collaborative knowledge exchange with the farmers, a quality that demonstrates a different role that businesses can play in farmers' development through communication. The buyers' close engagement with the farmers is driven by their own business needs and goals, such as the need of a coffee roaster and owner of a coffee shop to accentuate the quality of his coffee and highlight the provenance of the product for the benefit of the consumer. At the same time, the buyers' need for value addition to the product translates into new knowledge for the farmers who find out more information about the preferences of the consumers in export markets. It also leads to a more collaborative information exchange driven by the buyers' need and desire to better understand and learn from the farmers' expertise. As one buyer explained: 'We visit the group more often now and organise tasting and cupping events so that we both understand each other's needs and offer each other new knowledge. We want to scratch beneath the surface, learn from their expertise and traditional methods' (Buyer C). The following quote from another buyer is another example of the two-way, dialogical knowledge transfer:

> We must be involved in a constant dialogue, for the benefit of both sides. My background and experience is in marketing, so I can share my ideas with them. But, the technical side of the pepper production and processing is their expertise. I'm learning from them details, such as the importance of the garbling process, which I can use to promote and sell the pepper to my customers (Buyer E). (see also Touri 2018)

Willingness to listen to and recognise the farmers' valuable knowledge and expertise is a crucial practice in any development programme, but it is often bypassed as power dynamics and lack of communication between experts and farmers results in the marginalisation of local indigenous knowledge. In this case, through their engagement with the farmers, buyers display a quality that can be critical for the farmers' sense of empowerment and that is often hard to find in institutional development programmes. Seen from another angle, and given the role of buyers in transferring geographical knowledge to potential consumers, this communication is also useful in contributing to the representation of farming communities, beyond the over-used narratives of poverty, as the following quote shows: 'We never go down the 'poverty' narrative ... we do not touch that in our communication' (Buyer B). (See also Touri 2016, 2018).

The case of knowledge sharing on marketing product biographies and geographical knowledge displays the environmental concerns that these buyers share with the farmers, reiterating the much wider range of non-monetary relations that underpin economic practices including a joint commitment to environmental justice. The following quote from a buyer is testament of this:

> Low yield and lack of demand means that the Wayanadan variety [of pepper] has almost become extinct, and farmers are replacing it with higher yielding hybrid varieties, without realising this is something unique; if we don't help them, there may be no future not just for the product but for the entire community (Buyer E). (see also Touri 2016)

In essence, communication and knowledge exchange on product biographies encapsulates a number of practices associated with development communication programmes: the identification of problems and solutions, the sharing of information about consumer needs and realities, and the acquisition of new negotiation skills by the farmers that can prove valuable in assisting farmers to plan for future crops and increase profitability. However, unlike many development initiatives in the field of agriculture, the farmer-buyer relations display a more collaborative effort that is based on mutual recognition of each other's knowledge and expertise, and offers an example of the dynamic relations and sometimes intangible processes that are inherent in empowerment and development. Here, development happens in the context of a community collaboration process that can produce long-term social change for the farmers as they strengthen their marketing and negotiation skills in the global food market. In this process, the dialogue and communication with the buyers enables them to generate new ideas and innovative outcomes.

Understanding and appreciating such relations becomes possible with a weak communication approach to these relations; it means paying closer attention to practices such as the simple act of listening and knowledge exchange and how such practices can lead to development outside the formal boundaries of institutionally driven programmes. In the context of business relations more specifically, while strong theory would approach development and poverty eradication as reconcilable with top-down profit maximising solutions, a weak theoretical approach is more attentive and adaptive to local conditions and helps unearth the various development opportunities that reside in those hidden relations that underpin the global food economy.

Market Relations and Intangible Development Outcomes

The importance of the less tangible rewards, such as self-esteem and pride, that organic and fair trade farmers can gain from their participation in these networks, have been touched upon by few studies so far (e.g. see Murray et al. 2003; Raynolds 2002). For instance, organic agriculture relies on and respects indigenous knowledge, allowing farmers to have more control over the production process. Moreover, organic farmers' organisations, collaborative research, and knowledge sharing enables them to build stronger community relationships and develop better and more beneficial relations with external institutions, such as exporters (see Kilcher 2007). Such relations can prove significant for increasing self-esteem and confidence, and for the farmers in Wayanad, their interaction with the buyers becomes a source of motivation and encouragement helping farmers deal with social constraints and psychological barriers that can prove damaging for their economic development and sustainability.

The previous section referred to the role of cultural embeddedness in farmers' marketing strategies, but other types of embeddedness of economic action are also at play in this case. More specifically, the concepts of political and structural embeddedness, introduced by Zukin and DiMaggio (1990), are valuable for understanding the various ways in which the work and development of the Wayanadan farmers is mediated by local social norms and institutions. Understanding the role of such norms and institutions is crucial for evaluating the significance of the farmers' relations with the buyers for their own development. According to Zukin and DiMaggio (1990), political embeddedness concerns the impact that a struggle for power between economic actors and non-market institutions, such as the state, can have on economic actors and decisions. With the term structural embeddedness, introduced originally by Granovetter (1985), they refer to the way business behaviour and economic exchange is also shaped by patterns of ongoing interpersonal relations. In the case of the farmers in Wayanad, and assuming their status as business actors, their local social and cultural context is a manifestation of the interplay of structural and political pressures (embeddedness). The interviews revealed a strained relationship between farmers and local institutions such as government and banks that reflects a mutual lack of trust. From the perspective of institutions, this is often based on a misleading perception about farmers' economic insecurity. One of the farmers explained:

> Government officials and banks do not think highly of the farmers, they do not take us seriously, and that is why the young people do not want to get involved. Why should we encourage our kids to become farmers if they are going to have the same treatment from the society like we did? (Farmer A, Group Interview 10). (see also Touri 2018)

Other farmers explained how such perceptions inevitably impact their work. For instance, not being able to get a bank loan meant that they could not afford to use specific farming methods or grow certain varieties of crops. They are also faced with a government paradox that has made it much easier for farmers to take bank loans for chemical fertilisers, and as one of the farmers mentioned, 'It is actually easier to get the chemicals than it is to get a prescription for sleeping pills' (Farmer A, Group Interview 1). The pressures of this type of political embeddedness are exacerbated by cultural and social norms and interpersonal relations that often take the form of marginalisation and ridicule by their peers, rooted in society's (mis) perception about the economic returns of organic farming and the social status of organic farmers in particular.

One of the farmers who had recently converted to organic farming said: 'My own brother thinks I am crazy because of the low yield I get, compared to non-organic farming. He grows coffee in tons while mine comes only in kilos. They [the family] wonder why I even bother...' (Farmer B, Group Interview 4). Another farmer added: 'The local community thinks that organic farmers are crazy, because they are not making as much money as the others. They think we are not progressing because we are not becoming wealthier and there is no change in our lifestyle...' (Farmer A, Group Interview 4) (see also Touri 2016). These perceptions reflect bigger cultural trends in Kerala, where social mobility is driven by a widespread consumerist culture and the need for social recognition through the acquisition of wealth and consumer durables (Abdulla 2014). A rather peculiar manifestation of the consequences of this consumerist culture is consumers' prioritisation of the appearance of products like fruit and vegetables instead of quality. The farmers explained that although many locals may not afford to buy organic products, those who can afford them would still 'go by the appearance of the [non-organic] products because organic products do not look as impressive as the non-organic ones' (Farmer B, Group Interview 3). Another farmer explained, 'We get feedback from market traders that customers prefer the good-looking vegetables and the non-organic ones … especially women, who do most of the shopping … they look for bright colours and nice shapes' (Farmer C, Group Interview 4). As a result, a sizeable amount of organic products can only be sold as conventional products and at low prices. This creates further financial burden and an even more unfavourable and unsupportive environment for organic and sustainable farming, with several farmers resorting to excessive use of fertilisers in order to get high profits quickly. As one of the farmers said, 'A lot of people are just after the money—even if they are killing themselves' (Farmer A, Group Interview 8). According to another farmer, such trends and behaviours among farmers reflected a different philosophy and work ethic, with organic farmers clearly prioritising their health and the environment over the higher profits (see also Touri 2016).

During the second round of fieldwork, in 2016, several of the farmers expressed their concerns that not only had the pressure not eased but it also threatened their sustainability, as more and more young people abandoned organic farming. They feared that certain local varieties would become extinct, especially if more farmers turned to cash crops. Some of them confessed that these challenges were the reason why they would discourage their own children from getting involved in agriculture. The political, cultural, and social context and pressures described above underline the need for social scientists to adopt an interpretivist approach to development and well-being in the context of business relations and the global food market, recognising that actors construct meaning and make sense of the world intersubjectively (Pavarala 2020). In the case of the Wayanadan farmers, this means that the value of their relationship with the foreign buyers needs to be understood and evaluated in conjunction with their local reality and the benefits they perceive to gain from these relations. The following quotes from the farmers are examples of the non-monetary value that they gain from the visits of the foreign buyers:

> No one from the local market or the local government ever comes here to find out what we do. When the foreigners come to visit, this makes me feel important ... they are coming from so far away to see my farm, and they tell me they like my farming techniques. They are quite inquisitive, so I'm also doing my best to improve the quality of my products (Farmer C, Group Interview 3).

> All these foreigners who have come here have confirmed that our agricultural methods are amongst the best and thanks to that the soil conditions have remained good. They hold our traditional farming in high esteem' (Farmer A, Group Interview 5). (see also Touri 2016, 2018)

The discussions with the farmers revealed a strong sense of appreciation that they felt they received from the buyers. It was also through these buyers that the farmers felt a connection with the distant consumers. Several of the farmers said they strongly believed that consumers in Europe and North America were environmentally conscious and sensitive about farmers' working conditions. One of the farmers noted,

> There are a lot of foreign buyers visiting my farm and I understand that there are a lot of people abroad appreciating chemical-free food. I know not everyone can afford to have organic food. Even in the foreign markets, the premium is not affordable to everyone, but they appreciate our natural and genuine practices (Farmer C, Group Interview 7). (see also Touri 2016).

The farmer-buyer relation reflects a certain level of closeness and trust, and these are elements that have been recognised as integral to the relational social capital that sustains economic transactions and contributes to community development (Bolino et al. 2002). These components also find expression in Sen's identification of the powerful systems of values and norms and the non-profit motives that capitalism is dependent on. For Sen (1999: 262–264), behavioural ethics and trusted relations are a crucial ingredient for market success and are also integral to the process of the market mechanism that leads to development. At the same time, the value and recognition that the farmers gain from these trusted relations, encapsulate what Pavarala (2020) describes as 'intangible' development outcomes that are often the products of communication. In this case, it is particularly the 'listening' side of the farmer-buyer communication that proves most empowering, as it is a communication function that encourages a sense of confidence and capacity for change (Quarry and Ramírez 2009). The intangible outcomes are encapsulated in the sense of appreciation and recognition the farmers receive, which incentivises them to continue with their farming. These are the unanticipated and unintended effects and less visible consequences of the close relations and communication that underpin the market exchanges between farmers and buyers. They are more complex, emergent, and processual outcomes that are hard to capture through quantitative measures and indicators; yet, they can prove significant in less obvious ways.

The significance of the farmers' relations with the foreign buyers and the intangible outcomes can also be found in the slow, long-term, and more 'tangible' changes, and as a farmer indicated, 'Even the presence of a foreigner in the area can actually generate interest and curiosity from our local society, about our work' (Farmer A, Group Interview 2). This is a positive development, as the words of another farmer show: 'We are not losing hope—we are sure there is future for organic products and the local market will grow. People will realise the benefits of organic products and will consume them more' (Farmer B, Group Interview 5) (see also Touri 2016). During the second round of focus groups that took place in September 2016, several farmers had started creating short videos to educate the public about organic farming practices. These videos were screened during meetings organised by the local community, and several farmers appeared hopeful and optimistic about raising awareness and building closer relations with each other and with the community. One of the farmers' groups within the cooperative also established their own small body for tea procurement and marketing. An encouraging sign was the involvement of members of the younger generation from the farmers' families.

Conclusion

The close relations and regular communication that the farmers in Wayanad maintain with their foreign buyers encapsulate the essence of the hidden social relations that sustain the global market. These relations are not happening at the margins of capitalism but at the heart of it since both parties, buyers and farmers, are profit-seeking entities. Yet, they are also enterprises that share social, ecological, and ethical concerns, reminding us that social relations are inseparable from material outcomes (Leyshon and Lee 2003; Sen 1999). More importantly, the significance of these relations as a development process could not become visible without following the principles of weak communication, and without the more focused and localised description of farmers' experience and their actual relations with buyers, or without paying closer attention to local conditions and demands. In this chapter, weak communication informed the analysis on two interrelated levels. On a first level, and from the researcher's point of view, it drove a search for non-capitalist practices and non-capitalist activity that sustains global food networks. It prompted an exploration of alternative, non-monetised practices and provided the conceptual space for a re-reading of relations between two economic actors—buyers and food growers—through the prism of diverse economies and through a grassroots approach to communication that involved listening to the voices of the people. More importantly, by calling for approaches that are more attentive and adaptive to local conditions, weak communication reinforced the need to listen to the reality and the voices of the marginalised communities. On a second level, the concept of weak communication is embodied in the face-to-face, interpersonal, and collaborative communication that characterises the business relations between buyers and farmers, and the development outcomes it can generate for the farming community. Through this type of communication, the two parties have developed a community collaboration that is based on knowledge exchange and sharing of ideas and a recognition of their interdependence for a more sustainable food system. On the one hand, a vital component of this collaboration is the buyers' willingness to listen and engage in a more horizontal communication process than many development agencies. The buyers' willingness to learn from the farmers' experience empowers farmers with confidence and self-esteem that can lead to positive change. In such cases, the quality of relationships that underpin market transactions proves critical. They become embedded in a development process that is not limited to economic growth and imitation of achievements from the developed countries but includes less tangible development outcomes and self-efficacy. On the other hand, this communicative collaboration is also beneficial for the buyers. They can gain new knowledge about the origins of specific crops, which creates value addition for their products and helps them boost customer loyalty in a competitive food market.

There are important lessons to be taught from the less formal side of market relations and the less visible interactions in the food commodity chain. First, the close relations that the farmers in Wayanad have built with their buyers shows that businesses have an important role to play as part of their day-to-day activities. To a certain extent this is an unintentional process that would fit under the role of businesses as a development 'tool' (Blowfield 2010; Blowfield and Dolan 2014), as human well-being is created through engagement in commerce and trade and through wealth and value creation. However, the focus on communication reveals the multidimensional nature of development encouraged by business. In the case of Organic Wayanad, the type of development created by the buyer-farmers trade relations was not just material and growth-driven but psychological and social, addressing the farmers' need for self-confidence and capacity to envision change (Alsop and Heinsohn 2005). Cases like this can challenge hegemonic discourses that view business' role in development from the stance of offering financial aid to pull communities out of poverty. At the same time, recognising the multidimensional role that the private sector can play in development can contribute to discourses of sustainable development, highlighting the inextricable links between the economic, social, and environmental dimensions.

Second, the development opportunities that reside in such trade relations speak to McAnany's concerns about the need to look for new solutions and creative ideas in development and social change outside development programmes. A new area for such ideas could be the world of business (McAnany 2012). Looking at various market activities and relations will also create opportunities to focus more closely on actual everyday practices of communication on a grassroots level and move beyond institutional efforts that rely on large aid packages. Finally, a shift away from business' role in poverty alleviation and economic growth and towards social and human well-being through grassroots approaches is another way to bring to the forefront the multiplicity of resources, processes, and relations that make up sustainable development. This requires a shift away from what Merry (2016) calls the 'seductions of quantification' in search for the tangible, measurable outcomes and indicators of sustainable development. One of the reasons why sustainable development indicators have gained currency is their perceived value neutrality and political impartiality. Approaching the role of business through economic indicators and outcomes agrees with this perception, but as the experience of the Wayanadan farmers showed, businesses can become part of a bigger collaborative community that contributes to sustainable development in less visible, yet crucial ways, bringing more abstract yet equally significant benefits.

References

Abdulla, M. P. (2014). Emigration and consumerism – A regional experience. *OSR Journal of Economics and Finance, 2*(4), 41–44.

Alsop, R., & Heinsohn, N. (2005). *Measuring empowerment in practice: Structuring analysis and framing indicators.* World Bank policy research workingpaper3510.http://siteresources.worldbank.org/INTEMPOWERMENT/Resources/41307_wps3510.pdf

Ansari, S., Munir, K., & Gregg, T. (2012). Impact at the 'bottom of the pyramid': The role of social Capital in Capability Development and Community Empowerment. *Journal of Management Studies, 49*(4), 813–842. https://doi.org/10.1111/j.1467-6486.2012.01042.x.

Bakewell-Stone, P., Lieblein, G., & Francis, C. (2008). Potentials for organic agriculture to sustain livelihoods in Tanzania. *International Journal of Agricultural Sustainability, 6*(1), 22–36. https://doi.org/10.3763/ijas.2007.0266.

Banerjee, A., & Duflo, E. (2007). The economic lives of the poor. *Journal of Economic Perspectives, 21*(1), 141–168. https://doi.org/10.1257/jep.21.1.141.

Banks, N., & Hulme, D. (2014). New development alternatives or business as usual with a new face? The transformative potential of new actors and alliances in development. *Third World Quarterly, 35*(1), 181–195. https://doi.org/10.1080/01436597.2014.868997.

Bellemare, M. F. (2014). Development bloat: How mission creep harms the poor. *Foreign Affairsm.* Available at https://www.foreignaffairs.com/articles/africa/2014-01-05/development-bloat. Accessed 27 March 2018.

Blowfield, M. (2010). Business, corporate responsibility and poverty reduction. In P. Utting & J. C. Marques (Eds.), *Corporate social responsibility and regulatory governance* (International Political Economy Series). London: Palgrave Macmillan.

Blowfield, M. (2012). Business and development: Making sense of business as a development agent. *Corporate Governance, 12*(4), 414–426.

Blowfield, M., & Dolan, C. S. (2008). Stewards of virtue? The ethical dilemma of CSR in African agriculture. *Development and Change, 39*(1), 1–23.

Blowfield, M., & Dolan, C. S. (2014). Business as a development agent: Evidence of possibility and improbability. *Third World Quarterly, 35*(1), 22–42. https://doi.org/10.1080/01436597.2013.868982.

Bolino, M. C., Turnley, W. H., & Bloodgood, J. M. (2002). Citizenship behavior and the creation of social capital. *Academy of Management Review, 27*(4), 505–522.

Bornstein, D., & Davis, S. (2010). *Social entrepreneurship: What everyone needs to know*. New York: Oxford University Press.

Brigg, M. (2001). Empowering NGOs: The microcredit movement through Foucault's notion of dispositif, alternatives: Global, local. *Political, 26*(3), 233–258. https://doi.org/10.1177/030437540102600301.

Brigg, M. (2006). Disciplining the development subject: Neoliberal power and governance through microcredit. In J. L. Fernando (Ed.), *Microfinance: Perils and prospects*. London/New York: Routledge.

Bruck, C. (2006, October 30). Millions for millions. *The New Yorker*.

Collier, P. (2007). *The bottom billion: Why the poorest countries are failing and what can be done about it*. New York: Oxford University Press.

Davies, I., & Crane, A. (2010). Corporate social responsibility in small-and medium-size enterprises: Investigating employee engagement in fair trade companies. *Business Ethics a European Review, 19*(2). https://doi.org/10.1111/j.1467-8608.2010.01586.x.

Davies, I. A., & Ryals, L. (2010). A stage model for transitioning to KAM. *Journal of Marketing Management, 25*(9–10), 1027–1048. https://doi.org/10.1362/026725709X479354.

Easterly, W. (2006). *The white Man's burden: Why the West's efforts to aid the rest have done so much ill and so little good*. Oxford: Oxford University Press.

Economist. (2009). *Froth at the bottom of the pyramid*. https://www.economist.com/node/14298996

Enghel, F. (2015). Towards a political economy of communication in development? *Nordicom Review, 36*(special issue), 11–12.

Escobar, A. (2001). *Encountering development: The making and unmaking of the third world*. Princeton: Princeton University Press.

Fernando, J. L. (2006). *Microfinance – Perils and prospects*. Abingdon: Routledge.

Franz, M., & Hassler, M. (2010). The value of commodity biographies: Integrating tribal farmers in India into a global organic agro-food network. *Area, 42*(1), 25–34.

Gereffi, G. (1999). International trade and industrial upgrading in the apparel commodity chain. *Journal of International Economics, 48*(1), 37–70.

Gibson-Graham, J. K. (2006). *A postcapitalist politics*. Minneapolis/London: University of Minnesota Press.

Granovetter, M. (1985). Economic action and social structure: The problem of embeddedness. *American Journal of Sociology, 91*(3), 481–510.

Hall, J., Matos, S., Sheehan, L., & Silvestre, B. (2012). Entrepreneurship and innovation at the base of the pyramid: A recipe for inclusive growth or social exclusion. *Journal of Management Studies, 49*(4), 785–812. https://doi.org/10.1111/j.1467-6486.2012.01044.x.

Hammond, K., Katz, T., Tran, J. T., & Walker, C. (2007). *The next 4 billion: Market size and business strategy at the base of the pyramid*. Washington, DC: World Resources Institute.

Hassler, M., & Franz, M. (2012). The bridging role of intermediaries in food production networks: Indian organic pepper in Germany. *Journal of Economic and Social Geography, 104*(1), 29–40.

Karnani, A. (2007). *Doing well by doing good. Case study: 'Fair & Lovely' whitening cream*. Ross School of Business Paper no. 1063. Available at SSRN: http://ssrn.com/abstract=958087

Kilcher, L. (2007). How organic agriculture contributes to sustainable development. *Journal of Agricultural Research in the Tropics and Subtropics, 89*(Supplement), 31–49.

Kolk, A., & van Tulder, R. (2006). Poverty alleviation as business strategy? Evaluating commitments of frontrunner multinational corporations. *World Development, 34*(5), 789–801.

Kuriyan, R., & Ray, I. (2008). Information and communication Technologies for Development: The bottom of the pyramid model in practice. *The Information Society, 24*(2), 93–104. https://doi.org/10.1080/01972240701883948.

Leyshon, A., & Lee, R. (2003). Introduction: Alternative economic geographies. In A. Leyshon & R. Lee (Eds.), *Alternative Economic Spaces* (pp. 1–26). London\Thousand Oaks: SAGE.

Lyon, S. (2006). Evaluating fair trade consumption: Politics, defetishization and producer participation. *International Journal of Consumer Studies, 30*(5), 452–464.

McAnany, E. (1980). The role of information in communication with the rural poor: Some reflections. In E. McAnany (Ed.), *Communications in the Rural Third World* (pp. 3–18). New York: Praeger.

McAnany, E. G. (2012). *Saving the world: A brief history of communication for development and social change*. Urbana: University of Illinois Press.

Merry, S. E. (2016). *The seductions of quantification: Measuring human rights, gender violence, and sex trafficking*. Chicago: University of Chicago Press.

Mugwisi, T. (2015). Communicating agricultural information for development: The role of the Media in Zimbabwe. *Libri, 65*(4), 281–299.

Murray, D., Raynolds, L. T., & Taylor, P. L. (2003). *One cup at a time: Poverty alleviation and fair trade in Latin America*. Manuscript prepared for the Ford Foundation. Available at https://cfat.colostate.edu/wp-content/uploads/sites/63/2009/09/One-Cup-at-a-Time.pdf/. Accessed 15 May 2017.

Nakasone, E., Torero, M., & Minten, B. (2014). The power of information: The ICT revolution in agricultural development. *Annual Review of Resource Economics, 6*(1), 533–550.

Nega, B., & Schneider, G. (2013, January 3–5). *Social entrepreneurship, microfinance, and economic development in Africa*. Paper presented at the annual meeting of the Association for Evolutionary Economics, Philadelphia.

Newell, P., & Frynas, J. G. (2007). Beyond CSR? Business, poverty and social justice: An introduction. *Third World Quarterly, 28*(4), 669–681.

Nicholls, A. (2006). Introduction. In A. Nicholls (Ed.), *Social entrepreneurship. New models of sustainable social change* (pp. 1–35). Oxford: Oxford University Press.

Pavarala, V. (2020). Intangible outcomes (of communication for social change). In J. Tacchi & T. Tufte (Eds.), *Communicating for Change, Concepts to Think*. Cham: Palgrave Macmillan.

Pearson, R. (2007). Beyond Women Workers: Gendering CSR. *Third World Quarterly, 28*(4), 731–749.

Prahalad, C. K. (2009). *The fortune at the bottom of the pyramid; eradicating poverty through profits*. New Jersey: Pearson FT Press.

Prieto, M., & Quinteros, C. (2004). Never the twain shall meet? Women's organisations and trade unions in the maquila industry in Central America. *Development in Practice, 14*(1–2), 149–157. https://doi.org/10.1080/0961452032000170712Prieto-Carro.

Quarry, W., & Ramírez, R. (2009). *Communication for another development: Listening before telling*. New York: Zed Books.

Rahnema, M. (1991). Global poverty: A Pauperising myth. *Interculture, 24*(2), 4–51.

Raynolds, L. T. (2002). Consumer/producer links in fair trade coffee networks. *Sociologia Ruralis, 42*(4), 404–424.

Reed, A. M., & Reed, D. (2009). Partnerships for development: Four models of business involvement. *Journal of Business Ethics, 90*(1), 3–37. http://www.jstor.org/stable/40295083.

Ronchi, L. (2002). *The impact of fair trade on producers and their organizations: A case study with Coocafé in Costa Rica*. Poverty Research Unit: University of Sussex, Brighton.

Roxas, S. K., & Ungson, G. R. (2011). From alleviation to eradication: A reassessment of modernization, market-based, and communitarian solutions to global poverty. *Poverty & Public Policy, 3*(2) Art 5. https://doi.org/10.2202/1944-2858.1068.

Roy, A. (2010). *Poverty capital microfinance and the making of development*. New York: Routledge.

Samers, M. (2005). The myopia of 'diverse economies', or a critique of the 'informal economy'. *Antipode, 37*(5), 875–886.

Schanbacher, W. (2010). *The politics of food: The global conflict between food security and food sovereignty*. Westport: Praeger.

Seelos, C., & Mair, J. (2017). *Innovation and scaling for impact: How effective social enterprises do it*. Stanford: Stanford University Press.

Sen, A. K. (1999). *Development as Freedom*. Oxford: Oxford University Press.

Smith, S., & Dolan, C. (2006). Ethical trade: What does it mean for women Workers in African Horticulture? In S. Barrientos & C. Dolan (Eds.), *Ethical sourcing in the global food system* (pp. 79–95). London: Earthscan.

Touri, M. (2016). Development Communication in Alternative Food Networks: empowering Indian farmers through market relations. *Journal of International Communication, 22*(2), 209–228. https://doi.org/10.1080/13216597.2016.1175366.

Touri, M. (2018). Development and communication in trade relations: New synergies in theory and practice. *Development in Practice, 28*(3), 388–399. https://doi.org/10.1080/09614524.2018.1432569.

UNDP (United Nations Development Programme). (2008). *Capacity Development: Empowering People and Institutions*. Available at: http://www.undp.org/publications/annualreport2008/pdf/IAR2008_ENG_low.pdf. Accessed 2 Oct 2017.

Vakulabharanam, V., & Motiram, S. (2007). The ethics of microfinance and cooperation. *Ethics and Economics, 5*(1), 1–10.

Zukin, S., & DiMaggio, P. (1990). Introduction. In S. Zukin & P. DiMaggio (Eds.), *Structures of capital*. Cambridge: Cambridge University Press.

CHAPTER 5

Communication and Development in Fair Trade Movements: Producers' Empowerment Through Representation and Voice

Introduction

This chapter continues the conversation on the importance of the discourses that foreground conceptions and practices of sustainable development in the context of the global food system and engages with the power of the North to control ideas of sustainable development in the South this time through the symbols, images, and stories that make up the public representations of producers in the Global South. The chapter focuses specifically on fair trade, as the most archetypal expression of a globalised Alternative Food Network (AFN) for improving the lives of producers and their communities in the South. As a movement, fair trade has generated a consistent flow of visual and textual imaginaries that act as a set of semiotics between producers and consumers and constitute an integral strategy in fair trade's practices (Goodman et al. 2014). At the same time, these imaginaries rely on stereotypical representations of producers that suppress their voices. This chapter argues that this silencing not only undermines the knowledge of farmers, it also constitutes another way in which powerful institutions control and shape discourses of sustainable development. On the one hand, these discourses promote an understanding of sustainable development as economic growth that is based on the economic dependency of the Southern farmers on consumers from the North. On the other, silencing the farmers' voices reinforces perceptions of indigenous knowledge as inferior. This is a form of

M. Touri, *Sustainable Development and Communication in Global Food Networks*, https://doi.org/10.1007/978-3-030-46119-5_5

oppression that is manifested in a two-fold manner. First, it excludes the farmers from contributing to discourses of sustainable development, and in this way, it prevents bottom-up constructions of sustainable development discourses. This exclusion is aggravated by the absence of farmers' voices in academic discussions about the significance and potential impact of fair trade's representations on the farmers themselves. Second, silencing their voices obscures the multiple dimensions of sustainable development, and the role that these representations may play in the sustainability of farming communities.

The chapter begins by presenting an overview of the dominant discursive constructions through which farmers in the Global South have been represented historically, focusing specifically on the case of fair trade. The review is followed by a discussion of some of the most widely cited critiques, concerning the fetishisation of farmers' livelihoods and the contribution of these discursive embodiments in the continuation of colonial relations between the North and the South. Building on these critiques, the chapter argues that, by seeking to persuade consumers to vote with their dollars, such representations not only contribute to neocolonial relations, but they also embody a very narrow understanding of development that is confined to economic growth. This is in line with Western development paradigms and specifically 'the neoliberal economic-growth-oriented and market driven model of development that dominates world development today' (Tufte 2017: 168) and resonates with a broader agenda of neoliberalism, reducing social change to entrepreneurship and income generation in a market-based system (Wilkins and Enghel 2013: 168; see also Dutta 2011; Escobar 1998). In the same way that development models are exclusionary of communities' voices undermining their development, fair trade's discursive embodiments also lead to social subordination by excluding the voices of the producers from their own representations. This proposition is developed further through the notions of voice and recognition and with empirical evidence from fieldwork with farmers in Wayanad that sheds light on the ways in which this exclusion and subordination is experienced by the farmers. The empirical evidence reveals the complex and multifaceted way in which representation and misrepresentation matters for the well-being and sustainability of these farming communities, shedding light on the importance of the local context and culture in perpetuating the silencing of farmers' voices and feeding into their distorted images.

Building on the farmers' own experience and their need to have their voices heard, the rest of the chapter presents the results of an innovative collaborative experiment that involved the use of participatory video and storytelling techniques through which the farmers created their own short videos and text-based stories. This chapter offers evidence of the significant ways in which the farmers' inclusion in the construction of their own representations can contribute to their well-being and sustainability in a non-monetary way. This includes the sense of community empowerment enabled by the actual process of video production, rather than the finished product, and which creates the capacity for real-life change. Following the principles of weak communication and weak theory's mantra 'reading for difference rather than dominance', the chapter aims to bring into visibility the representation of farmers as another communicative practice that sustains the global food economy and can also lead to less visible but significant development outcomes.

Farmers' Representation in Fair Trade Movements

Fair trade has received considerable scholarly attention as a movement that is dedicated to the improvement of food growers' lives in the Global South and the creation of greater transparency around the conditions of production that encourages more ethical consumption. A critical aspect of this tenet lies in the shortening of the distance between producers and consumers through fair trade's moral education that is believed to encourage a social connection and consumer obligation towards the Southern producers (Goodman 2004; Low and Davenport 2005; Polynczuk-Alenius 2018; Shreck 2005). This moral education normally takes place through textual and visual narratives and imaginaries that are designed to offer a glimpse into the lives of the producers and become the vehicle through which consumer-producer solidarity is enacted. The discursive spaces through which producer and consumers are meant to get closer to each other are more than just marketing strategies. They are deeply inherent in the ethical development imperative of fair trade. It is through these semiotic connections that consumers will be encouraged to act at a distance and generate the required sales that will ultimately translate into economic and social support and development for the distant producers. As Goodman et al. (2014: 210) attest, without the discourses and images surrounding fair trade, consumers would be hard pressed to understand the difference that fair trade can make.

A rich body of ethnographic studies have cast light on the knowledges and discourses that travel within food networks, demonstrating aspects of the production process and origins of food, some of which are explored in more detail in Chap. 6 (Appadurai 1986; Cook and Crang 1996; Hughes and Reimer 2004; Jackson 2010). More importantly, it is through these knowledges and imaginaries that consumers are encouraged to act and get to 'see' the effect of their actions.

Fair trade's initial attempts for semiotic transparency included the narration of producers' lives through vignettes that Goodman et al. (2014) describe as the 'political ecological imaginary'. In this case, solidarity with consumers is created through visual and textual representations that focus on short descriptions of farmers' lives, emphasising their willingness to work the land and their gratitude to fair trade for enabling them to overcome the arduousness of the free market. These are often embellished with close-up smiling photographic portraits of the farmers working the land and with farmers' testimonies about the positive changes that fair trade brought to their lives, such as the capacity to buy clothes for the children, build a house, or send the children to school (Adams and Raisborough 2011; Wright 2004).

As fair trade gradually moved into the mainstream, the move marked an attempt to correct its 'poor taste' image and was followed by a shift in the marketing of fair trade products as quality branded products. Hence, images of hard-working and cheerful farmers were replaced by 'premium looking' images of spectacular landscapes and optimal growing locations (Goodman et al. 2014; Wright 2004). This strategy is also exemplified in a newspaper advertising campaign for Cafédirect, published in the Guardian newspaper between 1999 and 2002, where most of the attention is paid to the landscape and locations where the Cafédirect farmers work, implying that most of the work is actually done by the weather and geography (Wright 2004). Here, the emphasis on social justice and solidarity is replaced by the preoccupation with geographical landscapes that Wright (2004) interprets as 'an appeal to the imagined consumer's cultural capital, whose enjoyment of travel and tourism will ensure their appreciation' (p. 668). The same touristic and 'premium' representation of farming and labour is also reflected in a series of full-page magazine advertisements launched by Divine Chocolate in 2006. The campaign featured images of female farmers who were members of a cooperative in Ghana. The distinctiveness and 'premium' quality of these images lies not just in the absence of labour but also in the fact that the women appear to be professional models dressed in glamorous African dress (Wilson 2011). Following the quality embodiment, the semiotics of fair trade progressed to campaigns that involved media celebrities and which cemented fair trade's position in the market as a branded product. The role of British band Coldplay's frontman Chris Martin as a fair trade ambassador enrolled by Oxfam is amongst the most memorable instances of fair trade's celebritisation. Referring to this phase as the 'mirror of consumption', Goodman et al. (2014) are essentially drawing attention to and raising questions about the capacity of fair trade's discursive embodiments to add transparency to the inequalities of the global food trade, since what these narratives really achieve is to allow us to 'reflect more of our own selves back to our selves' (p. 217).

The celebritisation phase was followed by the so-called connected labour phase that finds expression particularly in the relaunch and repackaging of Cafédirect's lines, featuring images that attempt to make the food growers knowable to consumers by bringing the labour of the farmer back. This is done in a rather abstract way, focusing on images of farming tools—hand clippers, woven baskets, hand scoops, shovels—rather than the farmers. As Goodman et al. (2014: 217) explain, such images attempt to bring the labour of the farmers back into the picture but in a sort of 'phantom' manner and while also maintaining the emphasis on quality through 'premium looks' visuals. Despite the abstractness of the depicted labour and the farmers' background role in these campaigns, the images of farmers and their communities are not abandoned. Instead, they return to the forefront this time in the virtual spaces of brands like Cafédirect and the Fairtrade Foundation websites. In this case, the internet provides the platform where the embodiment of quality is combined with representations of the producer and his livelihood through images and farmers' own testimonials. Producer testimonies are used specifically as evidence of the various material ways in which food growers can benefit from fair trade premium and of the positive changes they achieved in their lives.

Polynczuk-Alenius and Pantti (2017) demonstrated how Pukka, another fair trade brand, employed visual and verbal representations that focus on the production processes, while keeping the farmer 'invisible'. This is similar to what Goodman et al. (2014) describe as the 'connected labour' phase that finds expression particularly in the packaging of Cafédirect's lines, featuring images that make the food growers knowable to consumers through an abstract portrayal of farmers' labour, focusing on images of farming tools rather than the farmers. Such images are often combined with an emphasis on product quality 'implied to be an outcome of fairness as embodied in the "labour of love" of the producers' (Adams and Raisborough 2011: 8.5). With fair trade moving into the mainstream, 'premium looking' images of spectacular landscapes and optimal growing locations became more prominent, implying that most of the work is done by the weather and geography, while the hands at work remain disembodied (Polynczuk-Alenius and Pantti 2017).

The thread that connects all these different embodiments is the tendency to avoid any overt discussion of development and encourage a rather individualistic and simplistic decision-making on behalf of the consumer that is based on self-reward. Fair trade's discursive embodiments fail to provide the pedagogical information that will bring the consumer closer to the social, economic, and environmental context of the producer. By doing so, such embodiments not only limit public engagement with issues of global development (Darnton and Kirk 2011; Tallon and McGregor 2014) but also become a source of oppression.

From Fetishism to Discursive Colonialism

Just as knowledge provisioning about fair trade moved to online spaces creating new possibilities for transparency, the same transparency simultaneously became murky as fair trade moved to supermarkets with its stories promoted through multinational corporations. Even if fair trade's virtual spaces allow for greater transparency, the social relations that underpin production remain concealed. Particular emphasis has been placed on the tendency of fair trade's representations to obscure the social, environmental, and economic dimensions of production and consumption, and its failure to challenge the structurally unequal status quo of free trade; and encourage a genuine sense of solidarity and cooperation between consumers and producers that is based on a common social and environmental cause rather than compassion and charity (Adams and Raisborough 2011; Dogra 2012; Hudson and Hudson 2003; Archer and Fritsch 2010; Clarke et al. 2007).

A central concern that emerges from this lack of transparency is the problem of commodity fetishism with farming communities and their livelihoods becoming romanticised and commoditised. Several aspects of the production—such as mechanisation—are omitted, while relationships between consumers and producers remain one-directional. The concept of commodity fetishism was developed by Karl Marx in the first volume of

Capital (1961), referring to the social relationships involved in commodity production as relationships between things rather than between people. Commodity fetishism reflects people's tendency to see their relationship with products as a relationship with money or a relationship between things rather than a relationship between people, that is, the producers. The focus remains on the characteristics of the final product and the value of products is seen as natural rather than social, with no consideration of the process of production (Hudson and Hudson 2003; Johnston 2008; Wright 2004).

For Hudson and Hudson (2003), the answer to fair trade's commodity fetishism lies in fulfilling its educational function through the provision of a genuine and reliable flow of information that will enable people to explore the social, ecological, and economic dimensions of food production. But as the previous section demonstrated, fair trade's discursive embodiments eventually contribute to further fetishisation that is masked by a carefully selective mediation of images, information, and knowledge (Lyon 2006; Wright 2004; Varul 2008). Although such knowledge is meant to add transparency to the production process, shorten the supply change, and bring consumers closer to producers, these representations are the output of cultural labour. As a result, the voices of producers belong to cultural intermediaries that inhibit the real participation of marginalised farming communities in their own representation (Adams and Raisborough 2011; Goodman et al. 2014); while at the same time reinforcing fair trade's role as 'a feel-good conscience cleanser for the well-to-do Northern consumer' (Hudson and Hudson 2003: 427). Consumers' role and their reaction to fair trade's discursive embodiments will be taken up in more detail in Chap. 6. What is most concerning here is the need for a more nuanced understanding of fair trade's underpinning discursive embodiments and the power of these embodiments to promote a de-politicised discourse of sustainable development by advancing a narrow and one-dimensional conception of poverty and development in the context of the global food trade.

Here, I am proposing a re-reading of these embodiments through the prism of development studies and humanitarian communication. To a certain extent, fair trade's representations reflect the discursive domination that the development industry encourages through texts and discourses that elevate certain forms of knowledge around the problematisation of poverty and frame marginalised communities as a homogenised mass to be uplifted by the developed world. Drawing on the work of Stuart Hall

(1997) and Assia Djebar (1992) on the unequal power relations between those constructing representations and those being represented, Manyozo uses the term 'spectacle of development' to describe the role of representation in the creation of colonialist discourses of difference. This spectacle '...alludes to the production, exchange and utilization of development imaginaries that are rooted in stereotypes, a failure to acknowledge difference and a disregard of the voices of subaltern groups' (Manyozo 2017: 14). At the heart of this idea lies Escobar's analysis of the 'discovery of poverty' that according to Manyozo, led to a spectacle of poverty and the imaginary and performance of the 'pornography of development' (Ibid: 15). A closer look at the images and representations deployed by fair trade reveals the promotion of a certain identification of farmers' lives and problems that is not based on the farmers' own local context and situation, but on the 'expert' professional discourses of institutions such as the Faitrade Foundation and their identification of problems, target groups, and solutions.

A better understanding of the reasons why such representations uphold a certain social order is possible through the notion of 'deliberate positivism' (Lidchi 1999; Chouliaraki 2010). 'Deliberate positivism' is often used with reference to imageries employed for humanitarian purposes that focus on the positive impact and the benefits of donations and fund-raising for the recipients of those funds. In a similar way to humanitarian campaigns, fair trade representations aim to provide evidence of the real lives of producers and, more importantly, of the improvements made to their lives thanks to the generosity of consumers in the North. Smiling faces and personal testimonies encourage an individualisation and personalisation that aims to emphasise the dignity and agency of the individual. More importantly, the role of these representations is to demonstrate, in a concrete manner, the positive change that audiences/consumers' actions can make (Chouliaraki 2010; Scott 2014). This results in the simplification and individualisation of the producer-consumer relationship with producers being portrayed as in need of support and encouragement, which they receive through consumer choice (Adams and Raisborough 2011). Beyond promoting a simplistic solution—that is, consumption—to a complex problem of inequality and development in the global food economy, these representations confirm and reinforce the superiority of the generous Northern consumer as the benevolent donors and saviours of the Southern producers (Cameron and Haanstra 2008; Chouliaraki 2010. As Wright (2004) posits, fair trade campaigns may try to celebrate the agency and rights of producers, but they also promote a reading of producers' lives and landscapes as 'consumables' through images of content farmers and personal stories that end up oversimplifying and dumbing down the idea of producer-consumer solidarity. Such a reading not only lacks reflexivity in terms of the power of consumption to create happiness and development but also suppresses the complexity of development per se and sustains clichéd neo-colonial images of the Global South as in need of salvation by the developed and advanced North (Hassan 2013; Wright 2004). In essence, implicit in these representations is a Western philosophy shaped by binary oppositions, that is, 'self'/'haves' (Northern consumers) and 'other'/'have nots' (Southern producers) (McEwan 2008: 122); and a modernisation logic based on which the sustainability of the Southern farming communities is tied to the economic assistance from the North.

The ways in which these simplistic representations impact on marginalised communities is a more complex issue. We can begin to understand this by following political economy approaches that shift the focus of attention towards the function of media representations within society and the ways in which the media reflect power imbalances between the North and the South (Scott 2014). Seen through the prism of political economy, dominant imaginaries and narratives of farmers and their livelihoods end up reinforcing these imbalances of power, affecting how the Global South is positioned in society and even the possibilities of development (Mezzana 2005 in Scott 2014: 183). If as Mezzana (ibid) attests, media representations of the Global South reveal the existence of various forms of colonialism, imperialism, and Western hegemony, similar forms are also reproduced and reinforced through fair trade. In this context although fair trade representations attempt to convey an optimistic image of the hard-working producers, they carry clear continuities with colonial representations, and reflect the reformulation of earlier narratives of late colonial enterprises that were based on the exploitation of labour legitimising the continuation of colonial rule (Wilson 2011). It is particularly the 'positivity' of images of hard-working, productive, and content farmers that frames 'empowerment' as the result of consumption and the market that eventually sustains a hierarchy of human life between the grateful receiver and the benevolent donor (Chouliaraki 2013). Not only do they reveal the existence of new forms of colonialism and Western hegemony, they also perpetuate imperialistic stereotypes of the Global South as inadequate and in need of the intervention of the powerful Northern consumer to achieve economic growth and modernity (Hassan 2013; Mezzana 2005). This is an ideological function of the media that is achieved through language and the circulation of discourses that relate to and shape systems of knowledge and social relations such as colonialism and dependency (Fairclough 1995). The work of Michel Foucault and Pierre Bourdieu on the significance of language in perpetuating certain power structures has been critical in informing development scholars' evaluations of the power of symbols to uphold a certain social order, the power of the First World over the Third. For instance, Manyozo explains that language becomes a source of oppression exercised through dismissal of the other's identity, through material and non-material violence, and through pity (Manyozo 2017: 37–44). At the core of this oppression lies the oppression of silencing the voices of the communities. In the next section I employ the concept of 'voice' to propose a more nuanced reading of fair trade's discursive embodiments and their significance for the farmers' well-being.

Representation as Recognition and the Power of Voice

In this section, I approach fair trade representations through theories of voice that encapsulate the practices of speaking and listening. Seen through this prism, the different waves of fair trade's representations are a typical example of what Bassel (2017) describes as patterns of audibility enforced by official discourses and institutions that prevent farmers from expressing their voice and being heard on their own terms. Instead, they promote and reinforce the already established image and position of the voiceless, stereotyped victim (pp. 18–19). Yet, a legitimate question was posed by Wright (2004): is partial misrepresentation and misrecognition the price to be paid for redistribution of wealth to producers, and is this a price worth paying? I explore this question through an analysis that combines a theoretical discussion of the politics of voice and empirical evidence from the focus groups with the farmers in Wayanad.

When the farmers in Wayanad were asked how they felt about fair trade's portrayals, their reactions were ambivalent at first. Some of them said they did not mind being portrayed as poor if it meant securing more customers; but most of them felt that this image was unfair and had to be rectified. It was specifically the discussion around the narratives of poverty that brought to light the sense of bitterness that farmers felt about how they were perceived by the world. More importantly, these narratives became the trigger for a deeper reflection of how they wanted to be perceived, demonstrating the value they saw in having their voices heard. As one of the farmers argued, 'Farmers play such an important role in society; everything we eat comes from the soil, you cannot make this with a PC, and still we do not receive the recognition we deserve' (Farmer E, Group Interview 16) (see also Touri 2020). Another participant added, 'People [consumers] just do not realise that they are the ones that need our support, not the other way round. Perceiving us as in need of aid is just wrong and needs to change' (Farmer B, Group Interview 13). This was a sentiment that was shared by most of the farmers that took part in the project, and although I made a point of clarifying that the narratives of fair trade were generic and were not always referring to a particular group or country, the majority of them reiterated the 'unfairness' of representing all the farmers of the Global South as a homogenous group.

For the farmers in Wayanad, being listened to becomes an issue of pride that reveals the need to be recognised and appreciated for their contribution to the environment and the world more broadly. In essence, their call for recognition on terms that deviate from the dominant narratives is a call for equal participation in the social interactions that underpin discursive constructions in the global food economy. The inequality that is inherent in these discursive constructions can be understood through theories that consider voice, recognition, and the ability to give an account of oneself as critical for democracy and social justice (Butler 2005; Couldry 2010). Voice can be conceptualised as a form of self-representation that allows individuals to participate in political life and be recognised. It is linked to issues of power, agency, and freedom to express an opinion, represent oneself, and be heard on one's own terms (Couldry 2010; Plush 2015). It is more than just speech and a vital element of human cultures, human development, and an essential communicative element of participative democracy (Milam and Heath 2014). Feminist scholars Rakow and Wackwitz (2004) also champion an understanding of voice as the means and ability to speak and to be taken into account in social and political life (p. 95). Access to or lack of voice can therefore produce specific relations of power by excluding the 'voiceless' from social groups and creating rage, antagonism, and embarrassment (Scruton 1983; Smith and Hyde 1991). However, it is not simply the 'access' to voice that matters, but the process of articulating one's voice. It is through this process that individuals can develop feelings of self-assurance and belief in their own identity. Such sentiments are critical for people's sense of connectedness with others (Podkalicka and Campbell 2010).

Yet, what should not be forgotten is that voice production requires resources such as media skills and literacy, which are distributed asymmetrically according to class, race, and gender, compromising the ability of marginalised communities to be heard by the powerful (Downing 2016; Madianou et al. 2015: 3022). At the same time, a redistribution of material resources would be inadequate without a shift in the hierarchies of value and esteem afforded different voices and identities (Dreher 2009). Value and esteem are inextricably linked to the importance of active listening and recognition, without which it is impossible to achieve interaction and connection among individuals and communities. For voice to be empowering, what matters is the quality of relationships between speakers and listeners, while valuing voice and prioritising listening entails recognition and respect for alternative forms of knowledge and responsive action (Couldry 2010; Honneth 1995; Tacchi 2011).

The work of Nancy Fraser is fruitful in unpicking the connection between representation and social subordination in the context of food networks. Fraser treats misrecognition as a question of social status and suggests that recognition should be conceived as a matter of justice that is inherently linked to the distribution of goods and rights. Her core argument is built on the nexus of distribution/maldistribution and recognition/misrecognition; but she is explicit in recognising that not all misrecognition is a side-effect of maldistribution. Instead, she argues that institutionalised patterns of cultural value are also responsible for impeding parity of participation in social life. Although class and status injustice is caused by the economic structure of capitalist society, status harms—such as misrecognition—that emerged as by-products of this economic structure have also developed a life of their own (Fraser and Honneth 2003: 23). If we conceive food network models such as fair trade as an institutional pattern of cultural value, farmers are denied the status of full participation in the social interactions that produce their representations. Fraser allows an understanding of the discursive constructions in food networks as an institutionalised relation of social subordination. To be misrepresented and misrecognised is a form of exclusion from the social interactions that shape food networks, and it renders farmers unworthy of respect or esteem. Fraser's theory legitimises the significance of representation in the structures of the global food economy as a source of oppression or potential empowerment for the farmers. This makes the equal participation of a misrecognised party in the construction of these narratives a vital component in the social justice remits of Alternative Food Networks.

Social Acceptability Through Voice: Overcoming Social Oppression

The focus groups with the farmers in Wayanad offered a valuable glimpse into how voice is thought about and how it emerges and is lived in the farmers' own cultural localities and meaning structures. Several farmers—especially those who had travelled to the Middle East and United States—explained they understood how these images reflected a marketing strategy, not just of corporations in the North but also of the regional middlemen who capitalised on the narratives of poverty and saving to increase their sales. As one of the participants explained,

> This is a shrewd strategy of the middlemen to paint us as living in poverty, for their own interests, so that they can sell more; also they can use us as a dumping ground for things they do not need, such as obsolete technology, fertilizers etc. (Farmer A, Group Interview 11)(see also Touri 2020).

Those farmers who were not aware of these narratives reacted with surprise, especially at the way corporations and even the Fairtrade Foundation would create such a uniform message. In the words of another participant, 'To us, fair trade is about 'fair' prices, it is not about eradicating poverty simply because we do not rely on fair trade for our income' (Farmer A, Group Interview 10); another one from the same group added, 'Yes, fair trade also tries to help the farmers from lower economic sectors; but this is a very one-dimensional image, it is a misrepresentation' (Farmer B, Group Interview 15).

Farmers think reflectively about sources of their 'oppression', and this became evident in their discussion about the role of local governments and organisations in perpetuating farmers' images as 'victims' and the implication of such portrayals. In the words of one of the farmers, 'It is in their interest to portray us poor and struggling, so they can secure funding from the state; and they do not really make any effort to understand the work we do' (Farmer C, Group Interview 14) (see also Touri 2020). As the discussions continued, all groups admitted that more than the image portrayed to the rest of the world, it was the local society's perceptions about farmers that was pernicious and had to change. They unanimously stated that people in Kerala itself have a skewed image of organic farmers, which is reinforced by the lack of respect and social acceptability they receive from official bodies such as local authorities and banks. This was even though, as several of the farmers admitted, their annual income could be higher than that of an employee. As one of the farmers explained,

> People think we are extremely poor and backwards. Even the local businesses and the church promote these images to the communities, and the people from the church come with charity handouts; but no one takes them because they do not need them. (Farmer D, Group Interview 10). (see also Touri 2018)

Another farmer also added:

> Organic farming is financially profitable and more sustainable as you can find natural fertilizers and pesticides within the farm itself. The land can give you, not just good and healthy food, but also medicines; and unless we protect all this, our future will be endangered (Farmer B, Group Interview 9).

Crucially, the farmers admitted that lack of recognition and the low social status of organic farmers, in particular (see Chap. 4), pose serious threats to the sustainability of the place, driving young people away from farming and from Wayanad. The farmers' sense of the inequality that exists in their local society is creating a need to make the value of their work recognised and respected. Through this discussion it became clear that being heard and making their presence felt, respected, and understood by powerful institutions was crucial. Some of them mentioned the need to get organised and become a force, as well as create a representative body that will gather information about farmers' rights. Others also highlighted the need for the different farmer groups in the consortium to have their own websites to establish their identity and make people aware of their work. Their need for recognition at the local level reinforces Bassel's (2017) argument about the importance of horizontal listening that is oriented to each other. According to Downing (2016), horizontal listening between peers will also increase the chances for marginalised voices to be heard and recognised by the powerful (p. 10).

The serious risks, which misrecognition poses to the sustainability of this farming community, bring to light a less visible path through which stereotypical representations of food growers matter, and which begins in their own local culture. Fraser's status model, that conceives misrecognition as status subordination, helps understand this social status injustice not as a by-product of maldistribution. Instead, it is the consequence of an institutional pattern of cultural value that privileges traits associated with specific professions (Fraser and Honneth 2003: 20). Ultimately, this shifts our attention to the underlying local power structures and inequalities that contribute to the construction of these narratives and from which the farmers are excluded as they are denied the space where their voice can be concretised. This raises important questions about the need to destabilise the traditional stereotypical narratives first at the local level and within the societal context of the farmers. It also means that we need to approach representation as a multilayered process that is not confined to the representation of unheard voices and creation of output—images and messages—that is communicated to the foreign consumer, but involves collective action that could lead to real-life change. Here, the connection between voice and space also becomes critical and as Mitra (2004) observes, when such space is denied in real life, a new space needs to be discovered and laid down. It was in search of this new space that a new communicative intervention was designed by the author and the farmers with the use of participatory video and digital storytelling tools.

Storytelling and Participatory Video as a Form of Empowerment

The focus groups with the farmers provided a space not just for data collection but also for a reflective discussion that allowed farmers to come together and reflect on their need to be recognised and achieve participatory parity in their community. In this way, these discussions informed our joint decision to conduct a storytelling experiment that would build a communication channel between the farmers in Wayanad and consumers in the Global North using the content and stories created by the farmers. Given the dearth of empirical studies that uncover farmers' own voices and undocumented narratives, and/or consumers' reactions to these voices,[1]

[1] Kleine's (2008) study has offered insight into the internet's role in connecting the spaces of production and consumption, but without addressing the feasibility of a more direct farmer-consumer communication, through narratives that have been created by food growers. See also the Fair Tracing Project (Light and Anderson 2009).

such an experiment could provide valuable insight into the feasibility of a more transparent communication between food growers and consumers across the globe. The consumers' responses will be discussed in Chap. 6. At this point, the focus is on the farmers' experience and degree of empowerment they felt through this form of voice articulation. For the farmers, the experiment presented an opportunity to 'get connected to like-minded people', as one of them put it. This was a chance for them to take a first step to correct the misleading narratives and educate consumers on issues around environmental sustainability. One of the participants explained the following:

> One of the main reasons why we are healthy and our land is prosperous is because of the good and natural food we are having and the natural methods we use; and people need to see this. They should see our farming methods, the actual process of organic farming, the results, the quality and size of the vegetables etc. It is only then they will understand why they are buying our products and how important farmers are (Farmer B, Group Interview 10).

For the more entrepreneurial farmers, this was an experiment for future farm tourism projects, especially since they strongly believed consumers should be able to have a first-hand experience of organic farming. More importantly, although not explicitly said, I sensed that for the farmers the experiment presented an opportunity to do exactly what they felt was necessary: mobilise themselves. For them, creating stories offered the departure point for something bigger—an opportunity to begin an ongoing dialogue with their own community and the world.

Storytelling

In the development sector, the articulation of alternative narratives is often exemplified through the process of storytelling, where telling one's story has potential effects at multiple levels of analysis, helping represent experiences and activate cognitive and emotional processing. Storytelling has been approached as an instance of voice (see Madianou et al. 2015) with its empowering nature stemming from providing a catalyst for self-reflection and community mobilisation in social and political life, enabling inclusion and empowerment through the experience of voicing one's own story, while it also enables a shared and scalable learning process in initiating future planning for social change. Through her ethnographic work in South Asia, Tacchi (2011) observed that practices of self-expression through storytelling can be particularly popular among communities, with the potential to contribute to social and political action. Along similar lines, the work of Cornwall (2016) portrays how film production becomes a means to collective empowerment and recognition for sex workers in India. Madianou et al. (2015) examined storytelling in the context of humanitarian digital technologies and their capacity to fulfil their potential for self-representation among communities affected by super-Typhoon Haiyan in the Philippines. Their findings indicated many instances where digital storytelling enabled participants to articulate voice through a range of activities. Storytelling can also become a 'soft' mechanism of learning that creates collective action and governance, as in the case of Antioquía in Peru, where stories were used by community members for the promotion of collaboration around tourism (Chapple and Montero 2016).

The power of storytelling stems from the fact that narrative is closer to lived experience in its appeal to multiple senses and is also easier to understand (Lambert 2013; Winskell and Enger 2014). Storytelling also provides a catalyst for self-reflection and community mobilisation in social and political life (Williams, et al. 2003; Rakow and Wackwitz 2004). In fact, storytelling is perceived both as social process and strategy of intervention in and outside the development sector. As a social process, collective storytelling enables inclusion and empowerment through the experience of voicing one's own story; while as intervention, it enables a shared and scalable learning process in initiating future planning for social change. Stories and narratives can be created through multiple media including audio, print, still images, and film and have the power to challenge our linear way of thinking (Servaes and Lie 2015). In the experiment conducted in Wayanad, the medium of choice was decided by the farmers who opted for short videos with the use of a mobile phone camera. Few of the farmers chose to write short stories embellished with still images. In total, the farmers created twenty videos and six written stories. Moreover, and following the farmers' suggestion, it was decided that the video production would be led by one of their peers who also worked as officer in the cooperative and would shoot and edit the videos. The video production took place between October and December 2016.

The Stories

Unsurprisingly, the farmers' videos and text-based stories presented a stark contrast to the narratives and representations described earlier. The average duration of the videos was 5 minutes and all videos and text stories focused on the farms, the different crops that grow in the area, as well as some farming methods used by the farmers in Wayanad. Several farmers talked about the importance of organic agriculture for the soil, the environment, and people's health, while several of them made a point in explaining why they converted from conventional to organic farming methods. In one of the text-based stories, a farmer called Francis described his very personal journey from using fertilisers to becoming an organic farmer. The switch to organic farming was caused by the tragic death of his 12-year-old daughter from cancer in the early 1990s. Since then, and after reading and researching about the use of fertilisers, he started experimenting with traditional seedlings and natural farming methods. Although initially his yield was low, his farm eventually became healthy and profitable. Francis' story is the only story that involves such a personal and emotional element. However, Francis' intention in the story is not to capitalise on the tragic event of his daughter's death and evoke an emotional response from the audience. His emphasis is on the process through which he converted to organic farming and the benefits of organic farming for the environment and people's health.

Most of the videos focused on the different crops growing in the area. While some farmers walked the viewer through their fields, others focused more closely on specific plants. These included the different varieties of pepper, the growing, harvesting and cleaning of coffee beans, bee farming, Ayurvedic medicinal plants, as well as the medicinal properties of turmeric and the 'Tulasi' plant (Holy Basil). Some of the videos also focused on specific types of organic compost, methods including irrigation, and the production of fertilisers, such as 'Jeevamrutham', a natural fertiliser made from local ingredients. In one video, the farmer explained and demonstrated the recipe for this specific fertiliser:

> To make Jeevamrutham, I need:
>
> 10 kg of cow dung from our Desi cow.
>
> 10 litres of cow urine, 2 kg of horse gram powder and 2 kg of jaggery.
>
> Then I take a bit of nutritious soil, I mix everything with 2 litres of water and I stir.

Every day I have to slowly stir the mix 3 times. I repeat this process for 4 days. I then dilute 1 litre of Jeevamrutham in 10 litres of water, and then I pour the diluted mix over the plants.

I have been using Jeevamrutham for the past five years and currently, my plants are healthy from disease and I get a good yield and premium quality products.

When I detect some disease on the crops or some unhealthy plants, I always use Jeevamrutham. I prepare the Jeevamrutham on Saturday and then it takes five days until I can pour it over the plants.

This is an extract from another text-based story:

> Initially my husband and I were practicing chemical farming, since 1998. But in the meantime, I read a lot about the side effects of chemical fertilisers and the damage they can cause to our health and the soil. I discussed this with my husband and we agreed to gradually start introducing organic fertilisers and other natural techniques. I also observed the practices of our neighbouring tribal community, the Kurichiya community, who were already practicing traditional organic methods and were enjoying a life free of diseases. This was a great motivation for me to start using their traditional practices. Once I started using organic methods, I watched closely as the soil and my plants slowly became healthier (see also Touri 2020).

All the videos and images were shot outdoors in the farms and most farmers were dressed in their everyday clothes. A few of them were featured wearing the traditional ‘lungi’, a wide strip of cloth used mostly by men, combined with a T-shirt or a shirt. Some of the male farmers were joined by their wives who are also their work partners in the cultivation of the land. In fact, the role of the family was mentioned in several of the videos and in the text-based stories, reflecting on the integral role of the family in farming and on the integral role of agriculture in the farmers’ personal life. This was succinctly summarised in the words of a farmer called Jayraman who said in his text-based story:

> Agriculture is part of our spirituality. This farm gives us pleasure, love and goodness. And we are giving the same back to the world! My most important motivation in farming is family. My wife Sheela will always be with me in our farm. We have two children, Revathi and Rahul, both are very interested in agriculture, especially in organic natural farming.

The use of technical language such as 'zero-budget farming', the description of their daily routines or of fertiliser recipes, and the portrayal of agriculture, not just as profession but as a vital part of their lives, contrast fair trade's heartfelt testimonies on several levels. Instead of focusing on moral competence and hard work, the stories of the Wayanadan farmers are based on more accurate portrayals of contemporary farming. The focus is shifted away from the image of the 'traditional', 'authentic', and 'charitable' producer to a more rational and realistic image of farmers as professional experts in their field. Moreover, the absence of romantic imagery encourages a connection and solidarity with the audience not through objectification but by portraying the farmers as competent professionals focusing on farming methods, recipes, and details about the uses of specific products in medicine. Many of the farmers also stressed their role and responsibility for 'feeding the world' and 'protecting the environment'. Besides, as Wright and Annes (2016) explain, symbols and narratives rooted in idealised imagery are unlikely to lead to a meaningful connection with consumers or produce an understanding of agriculture and rural life. The way the information is disseminated in these stories not only challenges the stereotypical representations promoted through the dominant discursive constructions of food networks; it can also offer a public service by educating consumers. A key question that emerges concerns the extent to which this type of stories and representations can truly encourage a reciprocal communication between farmers and consumers. I will return to this question in Chap. 6.

Participatory Video: The Power of Video Creation and the Playback Process

The video creation process proved to be a catalysing aspect of this experiment as it encouraged farmers' engagement with each other and led to further action. This was facilitated by the participatory nature of the video creation since at this stage the farmers took control and ownership of the project. Participatory video falls under the umbrella of participatory communication and participatory media approaches that endeavour to achieve a bottom-up communication and engagement with communities. Participatory video was first used in the 1970s in the National Film Board of Canada's Challenge for Change programme. The programme was led by community development worker Donald Snowden and engaged with

Fogo Islanders in Newfoundland. It led to the development of a new process of consultation and participation by communities that was later integrated in many development programmes led by NGOs, development practitioners, and communities in order to instigate social change and empowerment through dialogue (Crocker 2003; Harris 2009). Originally inspired by Paulo Freire's critical pedagogy of teaching and learning, the use of participatory video by development practitioners is aimed at catalysing collective action and building stronger relations between marginalised communities and external agencies (Shaw 2015).

In the context of global food networks, the value of participatory video lies beyond just representing the unheard voices of food growers. Instead, it can result in real-life change by engaging participants in a relational process towards community development (Shaw 2015). Participatory media, such as video, enable individuals to express themselves through their lived experiences and reconstruct the self-portrait of their community and culture. In other words, the process of video production, rather than the finished product, becomes central to community empowerment. This is line with the sense of empowerment associated with the production side of voice. In this case, engaging in the process of video production allows individuals—especially marginalised communities—to disrupt established narratives enforced by powerful institutional sources and reconstruct their identity based on new signs and codes (Rodriguez 2001: 763). Through this dynamic process of re-codification, participants can engage in dialogue with each other and with the community.

For the farmers in Wayanad, the participation in the process brought a new perspective, giving them a sense of pride and ownership (Bery 2003). To almost every one of the farmers who participated, the process of making the videos gave a renewed sense of confidence and trust in their work. There was a common feeling of 'being important' and a strong sense of pride that emerged from their answers which was connected to their capacity to express their voice through those videos. This is exemplified in the following quotes from the farmers:

> Creating the videos and talking to an imagined audience made me feel that I have a voice and I love talking about what I do [and] my farming methods (Farmer A, Group Interview 18).

> I'm very proud to have taken part in this [project] and I feel important (Farmer B, Group Interview 18) (see also Touri 2020).

At the same time, the production process can offer opportunities for participants to interact and reflect on their experiences together, and as the farmers in Wayanad explained, creating those videos enabled them to hear each other and talk to each other more than before.

The empowering nature of this interaction can be further illustrated through the video playback process. Watching footage collectively can be an intimate process, and playback provides the space for interaction between producers and viewers, it facilitates a feedback loop and a reflective self-appraisal that can build self-esteem (Lunch and Lunch 2006). In the case of the Wayanadan farmers, after the video creation was completed, a team of farmers and officials from the cooperative organised a local event where the videos were shared with farmers who participated in the project and with an audience of up to 50 farmers who had not taken part in the videos. The playback process and the opportunity to see themselves and hear their own voices on screen proved to be a nourishing and inspiring experience not just for the farmers who were featured in the videos but for the entire audience. Several of the farmers who had not taken part in the project commented on the sense of pride they felt to watch some of their own people on screen, and the informational value of these videos, as they learnt about new farming techniques and crops. Farmers were able to internalise the story and gain the power to take control, consider future action and local interventions that can bring positive change, and improve their quality of life (Bery 2003: 113). The shared experience of playback generated debate and new ideas about creating more videos for farmer training purposes and for educating local audiences about the benefits and value of organic farming. In this case, the participants' newly discovered ability to educate and encourage others to make positive change generated a sense of agency (see Bau 2014). This led to further action, and in September 2017, a group of farmers decided to organise screening events in neighbouring farming communities in Kerala and in Karnataka to showcase their work and their products to non-organic farmers and to local consumers. The events attracted audiences of up to 150 people and proved instrumental in improving social inclusion for the farmers through redressing some misconceptions about organic farmers' economic and social status. After the events, one officer from Organic Wayanad described how members of the audience were surprised with the variety of organic products that were cultivated in Wayanad and the fact that organic farming could be profitable. One farmer said: 'For a long time the locals laughed at us for caring about the environment; but they have now started appreciating our work. These videos are very important for the future generation' (Farmer, Individual interview). Another one also said, 'I feel more recognised after doing these videos, more people know us now and it helps farmers feel less isolated' (Farmer, Individual interview). Another group of farmers were motivated to create their own videos for marketing purposes. The videos were shared through the cooperative's social media account and their own YouTube channel, while some of these videos were also promoted by the European buyers through their own newsletters. These activities triggered iterative cycles of shooting and screening as the farmers began to experiment with filming techniques and playing back their videos to each other. Through these cycles, the video-making process created additional dialogue in response to changes that occurred. Specifically, the videos generated interest among fellow farmers and their foreign buyers, which led to further discussions, mobilisation, and creation of tailor-made videos for specific products—for example, turmeric—and for specific buyers. These strengthened the farmers' negotiation position as according to the Chairman of the cooperative 'Current buyers trust the farmers more now, and put fewer clauses in their contracts' (Chairman, Individual interview).

In this instance, participation in their own content creation enabled a sense of cohesion and solidarity in moving forward and initiating positive change at the local level. It also enabled a process resembling Paulo Freire's (2001) idea of 'praxis': a circular process of transformation involving critical reflection of one's reality followed by action and further reflection (Image 5.1).

The farmers' experience from the storytelling project confirms Manyozo's suggestion that voice should be a lived experience through which people question their socio-political reality (Manyozo 2016). Moreover, the capacity of voice to lead to change in the context of food networks needs to be understood within the global *and* local structural inequalities that shape farmers' lives. For the farmers in Wayanad, voice through storytelling and participatory video becomes a communicative pathway to question and influence the local context in which they live. Realising the value of their own videos as a tool for changing society's perceptions is an essential aspect of their empowerment. It is also a tool for them to begin challenging the dominant narratives of poverty at the grassroots level, through the production of alternative images and knowledges. A process that resembles Freire's *conscientização* and empowerment through dialogue is also taking place. Although the farmers displayed

Image 5.1 Farmers' videos shown during screening events

human agency from the start, in their awareness of the social, political, and economic contradictions, the storytelling process offered them a space for emancipatory dialogue that led to a renewed consciousness and appreciation of their capacity to take action. Here, the power of participatory video to bring everyone at the same level and create a sense of joint ownership and responsibility was a vital element in strengthening personal and group confidence and making the impossible look possible (Lunch and Lunch 2006).

Embedded in the video creation and playback process, and in the screening events, is also a listening process that involved fellow farmers and local consumers who showed promising signs of recognition and appreciation of the farmers' work. This creates the possibility of what Dobson (2014) describes as 'apophatic' listening that is key to effective dialogue as the listener abandons pre-established prejudice and makes a genuine effort to understand the speaker's reality (p. 178). For the farmers in Wayanad, the value of being listened to by farmers and consumers in their own community is best captured by philosopher Hannah Arendt's position about the importance of others' perception of us and the ways this perception affects how we can be present in the social and political realm (see Bickford 1996). The significance of this recognition lies in the way it gives farmers a sense of social connectedness, motivation, and renewed confidence to present themselves in their own community and deal with society's pressures. It is only in the context of the farmers' own societal backdrop (see Chap. 4) that one can comprehend the value they draw from being listened to and understood. It is for this reason that engaging with other local farmers and consumers becomes paramount for the Wayanadan farmers. As one of the participants explained, 'There is not enough awareness and information about the amount of care and effort that needs to go into organic farming, most people want to have good products for a cheap price' (Farmer D, Group Interview 27) (see also Touri 2020).

The Wayanadan farmers' experiment raises several questions about the feasibility of creating dialogical spaces in the context of global food networks where active listening has potential to occur and farmers' stereotypical representations can be challenged. For instance, are consumers—local or foreign—the right audiences for farmers' generated content and stories? Which other actors and institutions should 'listen to' farmers' representations? How could they offer attention, recognition, and response? As it has emerged from this research and the ongoing conversations I have had with the coordinators of the farmers' group, the foreign trading organisations that work closely with the farmers are already playing a vital role as cultural intermediaries. Through screening events, marketing material, and feedback loops, they are gradually creating channels for farmers' stories, knowledge, and images to reach a wider audience of influencers and consumers across Europe. It remains to be seen if these interconnections have the power to change public representations of producers in the Global South.

Conclusion

Fair trade's narratives have attracted criticism for perpetuating neo-colonial relations through representations that romanticise and fetishise farmers and the food production process. The potential implications these representations may have on farmers' own well-being have also raised questions about the perniciousness of such narratives, as they reinforce colonial relations and encourage simplistic pathways of intervention. The aim of this chapter was to begin complicating these narratives and unpacking the significance of representation for the process of sustainable development, through a theoretical discussion of representation as voice and a collaborative storytelling experiment with the farmers in Wayanad. The storytelling experiment gave farmers an opportunity to produce a series of short videos and text-based stories that not only demonstrated the capacity of farming communities to create their own representations but also provided a more realistic and accurate image of food growers as professional experts and environmental activists, rather than as 'traditional', 'authentic', and 'charitable'. More importantly, the storytelling experiment brings to light the multidimensional and complicated nature of representation in global food networks and the possible ways in which representation is important for the sustainability of the Southern producers.

First, for the farmers in Wayanad, there is an institutionalised subordination and violation of justice that begins in the farmers' own local society and, according to the farmers' testimonies, feeds into the construction of stereotypical representations of farmers. More than their connection to the distant Northern audiences, it is their local community and local reality that presents the main priority as they strive to achieve the recognition they deserve as respectable entrepreneurs and environmental activists. This draws attention also to the need for fair trade communication strategies to pay closer attention to farmers' local reality. Fair trade communication strategies are exclusively focused on creating connections between the 'poor' [farmer] and the 'powerful' [consumer], but the Wayanadan farmers' experience and need for recognition by their own community shows that a more horizontal communication and listening can be equally, if not more, crucial.

Second, the effect and value of farmers' representation for their lives is far more complicated than the response that farmers' images elicit from consumers. The process of self-representation and video-making can offer farmers a space for reflexivity, where they can negotiate the challenges posed by their local society and begin to develop strategies for action and real-life change. This offers a different approach to the importance and meaning of representation, which stems from the actual process of farmers' participation in the construction of their own narratives, rather than from the final output and the value attached to it by the audience. In this case, self-representation becomes an empowering lived experience and an integral component of fair trade movements that should not serve only to activate consumer purchasing decisions. Instead, it can enhance several human development aspects, beyond income generation—social freedoms, dignity, respect, belonging, and participation—that are critical for the social and economic sustainability of communities. Farmers' participation in their self-representation requires dialogue between fair trade organisations and the farming communities. Here, the role of intermediaries, such as buyers or coffee roasters can also prove instrumental. Dialogue presupposes that farmers are given the space to articulate their voice. Milam and Heath (2014) remind us that voice includes access to information and technical assistance to actively participate in decision-making, and this is exactly the type of support that fair trade organisations can provide so that farmers can actively participate in decisions about their own representations, locally and abroad. Fostering farmers' access to discussions and decisions concerning fair trade discourses and representations is also one way towards more democratic food networks. Although it may not lead to immediate change, it is a move that can prove critical for greater equality and justice in the structures of food networks. Creating opportunities and space for emancipatory dialogue can also give farmers a sense of enhanced self-esteem and a renewed consciousness and appreciation of their capacity to take action that is important for their well-being and sustainability. For farming communities in places like India, achieving recognition requires an organised and supported effort and constitutes a critical aspect of their development. In this case, self-representation can help primarily cultivate those skills that will enable them to take more organised action.

Third, the experience of the farmers in Wayanad brings to the fore the less visible ways in which communication matters for sustainable development. The farmers' desire to educate and share their specialised knowledge with audiences at home and abroad becomes another source of empowerment as it presents an opportunity to embark on a collective journey bringing together fellow farmers and consumers who share complementary interests in the environment and organic agriculture. The video creation experiment and the construction of their own images and stories constitute examples of weak communication as they had unanticipated yet powerful effects on the farmers, on an individual and collective level. These are instances of development and well-being that are not predicted by strong theories nor are they planned by development agencies.

Finally, the storytelling experiment revealed the significance of farmers' representation for deconstructing the idea of sustainable development. The critical role that self-representation plays in the farmers' sustainability reinforces the idea that sustainable development and social change is a highly nuanced process that is contingent on people's everyday realities rather than just a fight against an abstract global industrial food system (Arce et al. 2015). Representation is amongst these everyday realities. It affects farmers' sustainability by becoming internalised, and influencing their self-identity, psychological resources, and capabilities; and by generating knowledge and creating awareness about the benefits of organic agriculture. On another level, farmers' representations—especially when promoted through influential institutions such as fair trade—are integral in the construction of public discourses of sustainable development. 'These discourses are important, not least because our knowledge of the world is constructed in a variety of ways, through experience, learning, memory and imagination' (McEwan 2008: 122). Farmers have the knowledge and expertise to contribute to the construction of new 'experiences', 'learning', and 'imagination(s)'.

References

Adams, M., & Raisborough, J. (2011). Encountering the Fairtrade farmer: Solidarity, stereotypes and the self-control ethos. *Papers on Social Representations, 20*, 8.1–8.21.

Appadurai, A. (1986). Introduction: Commodities and the politics of value. In A. Appadurai (Ed.), *The social life of things: Commodities in Cultural Perspective* (pp. 3–63). Cambridge: Cambridge University Press.

Arce, A., Sherwood, S., & Paredes, M. (2015). Introduction: Food embodiments, assemblages and intersubjectivities: Ebbs and flows of critical food studies. In S. Sherwood, A. Arce, & M. Paredes (Eds.), *Food, agriculture and social change. The everyday vitality of Latin America* (pp. 1–20). New York: Routledge.

Archer, C., & Fritsch, S. (2010). Global fair trade: Humanizing globalization and reintroducing the normative to international political economy. *Review of International Political Economy, 17*(1), 103–128.

Bassel, L. (2017). *The politics of listening: Possibilities and challenges for democratic life* (Series: Palgrave pivot). London: Palgrave Macmillan.

Bau, V. (2014). Building peace through social change communication: Participatory video in conflict affected communities. *Community Development Journal, 50*(1), 121–137. https://doi.org/10.1093/cdj/bsu025.

Bery, R. (2003). Participatory video that empowers. In S. A. White (Ed.), *Participatory video: Images that transform and empower* (pp. 102–121). New Dehli: SAGE Publications Ltd..

Bickford, S. (1996). *The dissonance of democracy: Listening, conflict and citizenship.* Cornell Ithaca/London: University Press.

Butler, J. (2005). *Giving an account of oneself.* New York: Fordham University Press.

Cameron, J., & Haanstra, A. (2008). Development made sexy: How it happened and what it means. *Third World Quarterly, 29*(8), 1475–1489.

Chapple, K., & Montero, S. (2016). From learning to fragile governance: Regional economic development in rural Peru. *Journal of Rural Studies, 44*, 143–152.

Chouliaraki, L. (2010). Post-humanitarianism: Humanitarian communication beyond a politics of pity. *International Journal of Cultural Studies, 13*(2), 107–126.

Chouliaraki, L. (2013). *The ironic spectator: Solidarity in the age of posthumanitarianism.* Cambridge: Polity.

Clarke, N., Barnett, C., Cloke, P., & Malpass, A. (2007). Globalising the consumer: Doing politics in an ethical register. *Political Geography, 26*(3), 231–249.

Cook, I., & Crang, P. (1996). The world on a plate: Culinary culture, displacement & geographical knowledges. *Journal of Material Culture, 1*(2), 131–153.

Cornwall, A. (2016). Women's empowerment: What works? *Journal of International Development, 28*(3), 342–359. https://doi.org/10.1002/jid.3210.

Couldry, N. (2010). *Why voice matters: Culture and politics after neoliberalism.* London: SAGE.

Crocker, S. (2003). The Fogo process: Participatory communication in a globalizing world. In S. White (Ed.), *Participatory video: Images that transform and empower* (pp. 122–141). London: Sage.

Darnton, A., & Kirk, M. (2011). *Finding frames: New ways to engage the UK public in global poverty.* London: Department for International Development, BOND.

Djebar, A. (1992). *Women of Algiers in their apartment.* Charlottesville: University Press of Virginia.

Dobson, A. (2014). *Listening for democracy.* Oxford: Oxford University Press.

Dogra, N. (2012). *Representations of global poverty: Aid, development and international NGOs.* London: I.B.Tauris.

Downing, J. D. H. (2016). Forward. In O. Hemer & T. Tufte (Eds.), *Voice and matter: Communication, development and the cultural return* (pp. 9–11). Göteborg: Nordicom.

Dreher, T. (2009). Listening across difference: Media and multiculturalism beyond the politics of voice. *Continuum, 23*, 445–458.

Dutta, M. (2011). *Communicating social change, structure, culture and agency.* New York/London: Routledge.

Escobar, A. (1998). *La invención del Tercer Mundo: Construcción y deconstrucción del desarrollo.* Barcelona/Santa Fé de Bogotá: Norma.

Fairclough, N. (1995). *Critical discourse analysis: The critical study of language.* London: Longman.

Fraser, N., & Honneth, A. (2003). *Redistribution or recognition? A political philosophical debate.* London/New York: Verso.

Freire, P. (2001). *Pedagogy of the oppressed.* New York: Continuum.

Goodman, M. K. (2004). Reading fair trade: Political ecological imaginary and the moral economy of fair trade foods. *Political Geography, 23*(7), 891–915. https://doi.org/10.1016/j.polgeo.2004.05.013.

Goodman, D., DuPuis, M. E., & Goodman, M. K. (2014). *Alternative food networks: Knowledge, practice and politics.* New York: Routledge.

Hall, S. (1997). *Representation: Cultural representations and signifying practices.* London: Sage Publications & Open University.

Harris, U. S. (2009). Transforming images: Reimagining women's work through participatory video. *Development in Practice, 19*(4–5), 538–549. https://doi.org/10.1080/09614520902866405.

Hassan, F. (2013). Fair-trade practices in contemporary Bangladeshi society: The case of Aarong. In J. Servaes (Ed.), *Sustainability, participation and culture in communication: Theory and praxis* (pp. 101–120). Intellect: Bristol.

Honneth, A. (1995). *The struggle for recognition: The moral grammar of social conflicts.* Cambridge: Polity Press.

Hudson, L., & Hudson, M. (2003). Removing the veil? Commodity fetishism, fair trade, and the environment. *Organization & Environment, 16*(4), 413–430.

Hughes, A., & Reimer, S. (Eds.). (2004). *Geographies of commodity chains.* London: Routledge.

Jackson, P. (2010). Food stories: consumption in an age of anxiety. *Cultural Geographies, 17*(2), 147–165. https://doi.org/10.1177/1474474010363844.

Johnston, J. (2008). The citizen–consumer hybrid: Ideological tensions and the case of whole foods market. *Theory and Society, 37*(3), 229–270. https://doi.org/10.1007/s11186-007-9058-5.

Kleine, D. (2008). Negotiating partnerships, understanding power: Doing action research on Chilean Fairtrade wine value chains. *The Geographical Journal, 174*(2), 109–123.

Lambert, J. (2013). *Digital storytelling capturing lives, creating community.* New York: Routledge.

Lidchi, H. (1999). Finding the right image: British development NGOs and the regulation of imagery. In T. Skelton & T. Allen (Eds.), *Culture and global change.* London: Routledge.

Light, A., & Anderson, T. (2009). Research project as boundary object: Negotiating the conceptual design of a tool for international development. In I. Wagner, L. Ciolfi, H. Tellioglu, & E. Balka (Eds.), *ECSCW 2009: Proceedings of the 11th European conference on computer-supported cooperative work* (pp. 21–42). London: Springer.

Low, W., & Davenport, E. (2005). Postcards from the edge: Maintaining the 'alternative' character of fair trade. *Sustainable Development, 13*(3), 143–153.

Lunch, N., & Lunch, C. (2006). Insights into participatory video: A handbook for the field. *Insight.*

Lyon, S. (2006). Evaluating fair trade consumption: Politics, defetishization and producer participation. *International Journal of Consumer Studies, 30*(5), 452–464.

Madianou, M., Longboa, L., & Ong, J. C. (2015). Finding a voice through humanitarian technologies? Communication technologies and participation in disaster recovery. *International Journal of Communication, 9*, 3020–3038.

Manyozo, L. (2016). The language and voice of the oppressed. In O. Hemer & T. Tufte (Eds.), *Voice & Matter: Communication, development and the cultural return.* Göteborg: Nordicom, University of Gothenburg.

Manyozo, L. (2017). *Communicating development with communities.* Abingdon/Oxon/New York: Routledge.

Marx, K. (1961). *Capital: A critique of political economy volume I.* London: Lawrence & Wishart.

McEwan, C. (2008). *Postcolonialism and development.* London/New York: Routledge.

Mezzana, D. (2005). A cancerous image: The causes of Africa's negative and reductive image. *African Societies, 4.* Available at http://cyber.law.harvard.edu/digitaldemocracy/mezzana.htm. Accessed 30 Mar 2017.

Milam, J., & Heath, R. G. (2014). Participative democracy and voice: Rethinking community collaboration beyond neutral structures. *Journal of Applied Communication Research, 42*(4), 366–386. https://doi.org/10.1080/00909882.2014.911944.

Mitra, A. (2004). Voices of the marginalized on the internet: Examples from a website for women of South Asia. *Journal of Communication, 54*(3), 492–510. https://doi.org/10.1111/j.1460-2466.2004.tb02641.x.

Plush, T. (2015). Participatory video and citizen voice – We've raised their voices: Is anyone listening? *Glocal Times* no. 22/23.

Podkalicka, A., & Campbell, C. (2010). Understanding digital storytelling: Individual "voice" and community building in youth media programs. *International Journal of media, technology and lifelong learning, 6*(2), 208–218.

Polynczuk-Alenius, K. (2018). Mediating the agency of distant others: Proper distance in fair trade communication on Facebook. *International Journal of Cultural Studies, 21*(2), 155–172.

Polynczuk-Alenius, K., & Pantti, M. (2017). Branded solidarity in fair trade communication on facebook. *Globalizations, 14*(1), 66–80. https://doi.org/10.1080/14747731.2016.1175099.

Rakow, L. F., & Wackwitz, L. A. (2004). Voice in feminist communication theory. In L. F. Rakow & L. A. Wackwitz (Eds.), *Feminist communication theory: Selections in context* (pp. 93–110). New Delhi/London/Thousand Oaks: Sage.

Rodriguez, C. (2001). *Fissures in the Mediascape: An international study of Citizens' Media*. Cresskill: Hampton Press.

Scott, M. (2014). *Media and development*. London: Zed Books.

Scruton, R. (1983). *The aesthetic understanding: Essays in philosophy of art and culture*. Manchester: Carcanet.

Servaes, J., & Lie, R. (2015). New challenges for communication for sustainable development and social change: A review essay. *Journal of Multicultural Discourses, 10*(1), 124–148. https://doi.org/10.1080/17447143.2014.982655.

Shaw, J. (2015). Re-grounding participatory video within community emergence towards social accountability. *Community Development Journal, 50*(4), 624–643.

Shreck, A. (2005). Resistance, redistribution, and power in the fair trade banana initiative. *Agriculture and Human Values, 22*(1), 17–29. https://doi.org/10.1007/s10460-004-7227-y.

Smith, C. R., & Hyde, M. J. (1991). Rethinking the "public": The role of emotion in being with others. *The Quarterly Journal of Speech, 77*, 446–466.

Tacchi, J. (2011). Open content creation: The issue of voice and the challenges of listening. *New Media & Society, 14*(4), 652–668.

Tallon, R. A. M., & McGregor, A. (2014). Pitying the third world: Towards more progressive emotional responses to development education in schools. *Third World Quarterly, 35*(8), 1406–1422. https://doi.org/10.1080/01436597.2014.946259.

Touri, M. (2018). Development and communication in trade relations: New synergies in theory and practice. *Development in Practice, 28*(3), 388–399. https://doi.org/10.1080/09614524.2018.1432569.

Touri, M. (2020), 'Representation, voice, and producer-consumer communication in fair trade movements: toward new empirical directions', Geoforum,117:103–113 doi.org/https://doi.org/10.1016/j.geoforum.2020.09.011

Tufte, T. (2017). *Communication and social change: A citizen perspective*. Cambridge: Polity.

Varul, M. Z. (2008). Consuming the Campesino. *Cultural Studies, 22*(5), 654–679.

Wilkins, K. G., & Enghel, F. (2013). The privatization of development through global communication industries: Living proof? *Media, Culture and Society, 35*(2), 165–181.

Williams, L., Labonte, R., & O'Brien, M. (2003). Empowering social action through narratives of identity and culture. *Health Promotion International, 18*(1), 33–40. https://doi.org/10.1093/heapro/18.1.33.

Wilson, K. (2011). 'Race', gender and neoliberalism: Changing visual representations in development. *Third World Quarterly, 32*(2), 315–331. https://doi.org/10.1080/01436597.2011.560471.

Winskell, K., & Enger, D. (2014). Storytelling for social change. In K. G. Wilkins, T. Tufte, & R. Obregon (Eds.), *Handbook of Development Communication and Social Change* (pp. 189–206). Chichester: John Wiley & Sons Ltd.

Wright, C. (2004). Consuming lives, consuming landscapes. *Journal of International Development, 16*(5), 665–680.

Wright, W., & Annes, A. (2016). Farm women and the empowerment potential in value-added agriculture. *Rural Sociology, 81*(4), 545–571.

CHAPTER 6

Producer-Consumer Communication Through Digital Media: Empowering the Consumer

Introduction

In this chapter the focus shifts to a different relation in the global food system, namely the producer-consumer interaction. Consumers' relation with food is a complex process that takes place in and blends the private and the public, the commercial and the political spheres. This complexity has been captured by theories and debates around the conflation of consumption and citizenship, which have also raised questions about the legitimacy of the idea of political participation through consumption. The aim of this chapter is to explore consumers' complex relationship with food through a communication perspective and specifically through their communication with the distant producers. I turn to an unexplored aspect of globalised food networks, namely the interactivity and transparency of producer-consumer relations using technology and the spaces of the internet (Goodman et al. 2014: 242). Drawing on the principles of communication and collaboration (Chap. 3), the chapter begins a conversation about how might a non-monetary exchange, which is based on knowledge and information sharing, drive a change in consumers' mind-sets, and if/how such change could lead to a more meaningful consumer participation in sustainable development. The discussion is informed by data collected through the storytelling experiment that was introduced in Chap. 5. In this chapter, attention shifts to the consumers' engagement with the farmers' stories and the informational value they drew from them.

M. Touri, *Sustainable Development and Communication in Global Food Networks*, https://doi.org/10.1007/978-3-030-46119-5_6

A central notion in this chapter is the notion of sustainable consumption, which is embodied in UN's sustainable development Goal 12: Responsible Consumption and Production, and focuses on the need to tackle unsustainable consumption and 'reduce our ecological footprint by changing the way we produce and consume goods and resources' (UNDP). The idea of sustainable consumption is geared towards addressing the demand side and the ways in which goods and services can be consumed without compromising the Earth's carrying capacity and intergenerational equity. It also presupposes that change should not be limited to the practical side of consumption but include a redefinition of deeply held values (McGregor 2002).

Consumers' contribution to sustainable development has been theorised in relation to the civic and political dimension of their purchasing choices and habits, but the conflation of the acts of consumption and citizenship remains highly contested. Questions are also raised about sustainable consumption being nothing more than a contradiction in terms. This has created the need to distinguish between consumption and consumerism, with the latter representing high levels of economic and material consumption as a way of life that takes over human experience, accelerating environmental degradation (Evans and Jackson 2008). In contrast, consumption can be approached as a sociological act with the capacity to communicate meanings and contribute to the social world (Jackson 2006). Yet, in order to understand the potential power of food consumers to make a greater social and political contribution to sustainability, a change of mind-set is required that shifts our language and thinking from the 'consumer' or 'consumer/citizen' of food as an individual that uses the market as a platform for political action and change, to collective efforts and collaborations between organisations, businesses producers, and consumers. Reframing consumers' role means approaching consumers as individuals who when given an opportunity will do what they can to shape their choices for the benefit of all and use their relationship with food as a space for reflexivity and decision-making that is not limited to food but expands to other areas of life. This change of language needs to take place on every level and concerns any organisation, institution, or business that works with food and 'talks' about consumers, and once again the role of social scientists is crucial in laying the groundwork. The need for this change of mind-set is aptly explained in the following quote from the Food Ethics Council:

> The language of 'the Consumer' is pervasive in our food system. Whether working for a big retailer or a grassroots NGO, we talk about people as Consumers of food. To some extent, that's understandable: the most basic interaction we all have with food is to consume it. But language matters. It carries unspoken values and norms within it which shape our behaviour as individuals, and which add up to Mindsets that in turn shape entire human systems ... everything from our motivation to participate in our local community, to our concern for others and the environment can be reduced by priming the Consumer Mindset. (Food Ethics Council 2017: 14)

Hence, sustainable consumption should not be understood as an individual choice of a sustainable lifestyle but as woven into social structures and actions that lie beyond economic and material consumption and can provide for simple pleasures such as energy saving or adopting healthier and environmentally friendlier eating habits (Evans and Jackson 2008; Trentmann 2007).

The focus of this chapter lies specifically on the role of communication, information sharing, and collaboration between food producers and consumers across the globe in enabling a more meaningful participation of food consumers in sustainable development. Until now, our knowledge surrounding the role of information tends to focus on public service and marketing campaigns, and labelling strategies that inform consumers' decision-making, but this model has proved insufficient in understanding the role of the consumer. Not only are there vast variations in consumer practices and decisions across the different countries, cultural contexts, and between social groups but also the information that is commonly available through labels, marketing material, and the media is not sufficient to bring about the changes that will contribute to sustainable development (Hinton and Goodman 2010; Reisch 2004). Instead, the often generic and impersonal information that is designed to engage consumers results mainly in alienating them from sustainable lifestyles (Hinton and Goodman 2010).

In response to this deficit, this chapter introduces a different model of information provision that draws on the active participation of food growers as information and content providers, and consumers' virtual participation in an interaction with the producers. At the core of this approach lies a shift in the way we think and talk about producers and consumers, treating both as 'fellow citizens, able to share concerns and build mutual understanding' (Food Ethics Council 2017: 23). In this case, the consumers' receptiveness and response to farmer-created content, which differs to the usual information and labelling strategies, have provided valuable insights in the different ways that a more direct and personal producer-consumer communication can inform consumer reflexivity and engagement with food in different ways. This is a type of collaborative communication that involves sharing of ideas, praise, and encouragement and an attempt to create a joint meaning about food and the environment (Kumpulainen and Kaartinen 2003).

The chapter begins with a review of key theories of consumer citizenship and political consumerism and their relevance to sustainable development. Attention is drawn particularly to the importance of information and knowledge sharing in activating processes of consumer reflexivity and citizen mobilisation for sustainable development. The role of the citizen consumer is also discussed in the context of AFNs, where the consumers' need for reconnection with the origins and production of their food is believed to be paramount for their mobilisation towards more conscious and sustainable lifestyles. The issue of reconnection is at the heart of several debates that concern the return to more localised and community-based food provision schemes as a form of resistance to the industrial globalised food system. Local food networks are also hailed for enabling closer producer-consumer relations and better information and knowledge sharing about the origins and production of food. However, an over-emphasis on localness and proximity carries the risk of defensive localism that undermines the global dimension of sustainability and the need for cosmopolitan approaches and solutions. Although this chapter is concerned with globalised food networks, the discussion on the local-global tensions in discourses around food networks is important for revealing the relationality of the local and the global.

The chapter then turns to the main communication channels through which consumers obtain knowledge that informs their actions in the fair trade and organic movement. The discussion focuses on the opportunities and limitations that exist for closer producer-consumer relations in global food networks and for a more active role for consumers, paving the way for the introduction of digital storytelling as a potential avenue for more horizontal, dynamic, and collaborative producer-consumer communication that could also encourage more reflexive and sustainable consumer attitudes. The remaining of the chapter builds on the findings from the focus groups that were conducted with consumers in Leicester and London in June and July 2017. Through the consumers' experience and engagement with the videos of the Wayanadan farmers, the chapter presents the new opportunities that a more interactive communication between producers and consumers can generate for consumer reflexivity, and ability to reconsider their behaviour and engage with the food production process independent of location. In line with the idea of the weak communication developed in this book, the chapter calls for the need to be more open and attentive to informal communicative practices, such as storytelling, which can inspire subtle but meaningful relations among individuals and communities and can encourage a more holistic understanding of sustainable development.

Consumer Mobilisation and Political Consumerism in Global Food Networks

We live in the era of McDonaldization (Ritzer 2003) where the consumption of signs, symbols, and products increasingly compensates for our individualistic society and lack of community. Modern life is dominated by ideas and processes of rationalisation, as well as by precariousness, instability, and decomposition of social forms, giving rise to what Bauman describes as a 'liquid' modernity (Bauman 2007). Liquid modernity negates political life as power is gradually being divorced from 'solid' politics, lifestyles, and consumption crossing over to the realm of political participation, while the consumer market place becomes a sphere of political and civic action. The conflation of citizens/voters and consumers, where consumers become surrogate citizens, is not a new phenomenon. Instances of so-called political consumerism or consumer citizenship can be found in practices as old as the 'made in' label that was used by the British government after the end of WWI as a warning sign to prevent consumers from buying German products (Van Ham 2010). Yet, it is through more recent developments including government privatisation policies and increased citizen disengagement from politics that consumer choices have increasingly acquired political character, with consumers playing a more central role in expressing notions of citizenship. As a term, 'political consumerism' became popularised during the 1990s, through Michele Michelleti's discussion of the boycott of oil company Shell in

Denmark as a form of political participation. The idea of political consumerism is a form of non-institutional and personalised action that encompasses elements of civic engagement and consumption, as individuals use the market as a platform for political action and change (Bennett 2003; Dalton 2008; Stolle et al. 2005). Political consumerism is inextricably linked with ethical purchasing, and although ethical consumption can exist outside the realm of politics, political consumerism comprises politicisation *through* ethical consumption (Lekakis 2013). Political consumers are concerned with the biography and power relations inherent in products and become active through boycotts, buycotts, discursive forms of political consumerism, lifestyle commitments, and communities (Boström et al. 2019).

Individual consumer practices have become central in emerging notions of citizenship also in the context of sustainable development, what Michelletti and Stolle (2012) refer to as 'sustainable citizenship'. In line with definitions of sustainable development that focus on practices that safeguard nature and promote social justice, sustainable citizenship encompasses a set of beliefs, practices, and lifestyles that people should embrace in order to promote economic, environmental, and social development. Similar to the notion of consumer citizenship, sustainable citizenship relies on reflexivity and individualised responsibility (ibid). Sustainable citizenship also reflects the need for rethinking one's rights and responsibilities in a global context for creating and governing a sustainable world (Kurian et al. 2014). Sustainable citizenship has been articulated with particular emphasis on notions of green and environmental citizenship (Whitehead and Bullen 2005; Smith and Pangsapa 2008). With relationships between national governments and transnational corporations becoming more fluid and evasive, consumers' role and responsibility in protecting fair economic distribution and the natural environment becomes more pressing (Klein 2010; Sassatelli 2007; Soper 2007; Wheeler 2012). Food is a significant arena in which consumers are engaging in politically motivated actions and choices, including traditional forms of collective organisation and more unorganised, private practices, such as shopping, demonstrating their social and environmental concerns (Gasparėnienė et al. 2017).

Of course, the conflation of citizenship and consumption raises critical questions about the unstable construction of citizenship that is defined through choice and a personalisation of collective action (Bennett and Segerberg 2012). Based on an analysis of coffee activism through fair trade, Lekakis (2013) asserts that the collectiveness required by meaningful politics and global social change renders political consumerism invalid. In their discussion of sustainable citizenship, Michelletti and Stolle (2012) also raise questions about the effectiveness of do-it-yourself individualised practices given the easiness with which people can withdraw from these practices, or the risk of free-riding when consumers expect others to commit to expectations for sustainable consumption. Nevertheless, the potential for people to use the market as a platform for political expression should not be entirely discounted. Not all consumption is depoliticised and many citizens *do* consider the consequences of their food choices and utilise Alternative Food Networks as a space to send signals to producers and institutions, although institutional guidance is required for achieving higher levels of politicised consumption (Barnett et al. 2005; Michelletti and Stolle 2012).

A powerful dimension of this type of individualised action lies in its capacity to activate processes of reflexivity in relation to the consequences of lifestyles and consumption choices on nature, humans, and animals. The power of these processes lies in the broadening of the spatial and temporal dimension of citizenship to include consideration of the global common good and the ways in which even past behaviours can affect the future. Besides, as Canclini (2001) argues, questions around our citizenship are often answered through our private consumption habits rather than through our participation in political organisations. Hence, consumption can turn into a site that is good for reflexivity and where a good part of economic, socio-political, and psychological rationality is organised (ibid: 5). Communication and information sharing can prove crucial in activating such processes of reflexivity, and the closer relations between producers and consumers can activate subtle types of sustainable citizenship and citizen mobilisation for sustainable development.

Alternative Food Networks are broadly perceived to empower consumers through knowledge acquisition of agricultural practices and their environmental and social consequences, engaging them in debates over environmental sustainability and social justice (Levkoe 2006). However, they also reflect a perspective of consumption as the channel through which we can 'buy our way' to a more equitable world (Micheletti 2003). Consumer mobilisation in AFNs can occur through involvement in campaigns against environmental degradation and genetically modified products or unfair trade relations, especially for smallholders in the Global South. Approaches to the mobilisation of the citizen consumer in AFNs are broadly divided in three areas: local networks or short food supply chains, organic networks, and fair trade movements. In all three cases, consumer actions are driven by a combination of self-interest and a sense of responsibility towards the common good, and in all three cases these actions are aided by the availability of information concerning product biographies and trade relations. Although the focus of this book is on globalised networks, discussions on local food networks are paramount for they provide insight in the relationality of the 'local' and the 'global'.

(Re)Connecting Producers and Consumers Across Space: The Challenges of Unreflexive Localism

Consumers' need for reconnection with their food has emerged from the growing geographical and cognitive distance that separates food production and consumption in the global food market, with production chains becoming more complex and less transparent (Hendrickson and Heffernan 2002; Eden et al. 2008). In essence, 'reconnection' presupposes 'disconnection' and 'disembeddedness' in the food system, an idea that became more prominent during the 1990s following food crises, such as BSE, and loss of agricultural diversity, identity, and environmental degradation (Dowler et al. 2010; Feagan 2007). It is in this context that scholars have advocated localised food systems as the answer to global industrial agriculture and a solution for re-establishing closer socio-ecological relationships between producers and consumers, securing more ecologically benign farming systems and supporting local economies (e.g. DuPuis and Goodman 2005; Lamine 2005; Smithers et al. 2008). Localised food systems are associated with 'socially embedded' economies that, in contrast to the globalised food system, encourage connections between producers and consumers, educate consumers about food origins, and strengthen local economies (Seyfang 2006; Whatmore and Thorne 1997).

One of the key advantages of localised food initiatives is perceived to be the material or biological connection encouraged through short food supply chains and community supported agriculture (CSA) schemes that provide space for regular, real-time connection between consumers and food (Fonte 2013). These connections can take place through face-to-face sales in farmers' markets and farm shops or box schemes. Community supported agriculture, an idea and practice that was first developed in Japan before it spread to the rest of the world, encapsulates many of the principles of producer-consumer reconnection. It is based on the mutual commitment between a farm and a community of supporters/consumers and underpinned by ideas of solidarity, fair price, and respect for the environment (IFOAM). Consumers that subscribe to such schemes receive a regular box of in-season fruits and vegetables and dried goods. More importantly, they are given the opportunity to become more actively involved in the production process and build closer relations with farmers through newsletters or open-farm events. Knowledge and information is commonly shared through newsletters, internet blogs, and meetings with shareholders. Participants are also offered the opportunity to attend workshops where they are educated about cultivation methods (Opitz et al. 2017). As Vogl et al. (2004) report, consumers' involvement in these activities can facilitate better understanding of the challenges producers have to cope with. They also develop greater appreciation for food and its origins, which is central to the social identity that consumers build through their involvement in AFNs (Jarosz 2008; Renting et al. 2003). Systems such as CSA capture the environmental and economic value of proximity through support for local economies and the reduction of energy use and greenhouse gas emissions. Concerns for climate change and environmental degradation have been a key source of motivation for re-localising food and a shift towards self-interest and a more self-contained and resilient future. Yet, such transitions also raise questions around issues of exclusivity and lack of engagement with greater social justice and environmental protection (Sage 2014).

AFNs literature and food activist discourse, both in Europe and United States, embrace the idea of localisation as a solution to the problems of global industrial agriculture, positioning 'local' food chains as the antidote to the environmental and economic detriments of the global food economy. This idea has attracted criticism by scholars who have drawn attention to the damaging consequences that 'unreflexive' localism can have on the social justice remit of Alternative Food Networks by leading to a 'defensive militant particularism' (DuPuis and Goodman 2005, p. 362). Hinrichs (2003) elaborates by explaining how localisation becomes a defensive mechanism that emphasises the opposition between the homogeneity and coherence of the 'local' to the heterogeneous and destabilising global 'other'. This was reflected in the attitudes of some of the consumers who took part in this research, as the following statements show:

> I'm more interested to know where it's [food] come from. I'd prefer it if it's not travelled so far… (Participant A, Group Interview 1)

> …I wouldn't buy eggs from Romania, I wouldn't fancy that! (Participant B, Group Interview 1)

> I always buy Scottish salmon, I don't know why … If it said something like 'Polish' salmon I'd give it a miss… (Participant E, Group Interview 7)

Here, environmental concerns such as food miles are blurred with a defensive attitude towards the 'foreign' product, but the practice of defensive localisation implies a restricted consumer role in the promotion of sustainability since it presupposes a connection between community building in localised production-consumption networks and resistance to capitalist relations, a connection that remains dubious (Lockie 2009). Moreover, citizenship is global and cosmopolitan; it requires education on the relativity of places and the de-territorialised ethic of care towards others and the environment (Ibid). Besides, food origins and provenance extends beyond spatial locality and encompasses an entire philosophy of food production. Rather than treating provenance as an objective attribute of the product, we should approach it as a negotiated quality that is accomplished through a combination of elements from soil and climate to elements such as the philosophy of the producer and regional traditions; and which is constructed by all actors along the production chain (Light 2014; Smith Maguire 2012: 370–371).

Drawing on the work of Harvey (2001), DuPuis and Goodman (2005) introduce the idea of 'unreflexive' politics to describe the authoritarian premise of localism. This authoritarianism stems from the narrow, elite interests that localism represents, where social change is premised on the utopian ideal of a small and non-representative group of people (Childs 2003). In this context, food consumers' activism is limited within spatially defined projects. The limitations of this approach become obvious with the aid of the work of scholars, including Childs (2003), Amin (2002), Massey (1994) and Castree (2004), who view place in a non-territorial manner, championing the idea of an open politics of reflexivity and respect. Amin's (2002: 391–397) conception of a politics 'in place' captures the essence of place as a dynamic set of nodes and moments in a network, while political activity becomes varied, plural, open, and contested, juxtaposing the local, national, and global.

Seen through the prism of a reflexive localisation, the role of the consumer in AFNs should therefore be based on awareness and understanding of the different ways in which locales and communities connect with one another across space and the importance of the linkages between distant markets for the sustainability of the food system. Producers and consumers are part of both the global and the local, and producers of specific crops are often dependent on export markets and international consumers/buyers for the sustainability of their farms. Despite their 'defensive' attitude, the consumers who were interviewed for this research were also conscious of this local-global interconnection, and as one of the participants commented: '...coffee, chocolate, black pepper, turmeric, aloe vera ... these are things you can't get over here, and these farmers need support to be able to continue growing them' (Participant C, Group Interview 3). It is important that consumers remain open to the possibility that the promotion of a transparent relationship with producers, the visibility of the point of food production, and the fostering of trust can actually materialise beyond the confines of the locality. It is this openness that renders consumer agency significant in food production networks, and which also requires education and information that can encourage unmediated relationships. Developing a broad knowledge about the way foodways operate in one's environment and beyond is one of the ways towards a sustainable food system (Booth and Coveney 2015: 18–19).

In the following sections, I engage with the different channels through which consumers are informed and mobilised in the two main types of global networks that this book has focused on so far, the fair trade and organic movements. In each case, consumers are invited to 'participate' and support food growers through the consumption of information-laden products. However, in both cases, the existing communication channels that are meant to facilitate closer producer-consumer connections and solidarity end up encouraging the construction of forms of consumer identity that fail to enact the reflexive questioning that should underpin the more politicised global citizen.

Consumer Activism Through Fair Trade Communication and Cosmopolitan Sympathy

In theory, fair trade consumer activism is in line with Micheletti and Stolle's idea of sustainable citizenship which 'holds the central claim that people should do all they possibly can to help improve social justice and safeguard nature to make the world a better place in which to live' (2012: 89). Fair trade's dominant citizenship expectation lies in developing lifestyles that do not endanger the life of future generations. Yet, Lekakis (2013) analyses fair trade's communication as an organisational narrative that fails to make a connection between the civic and consumer duty or encourage consumer involvement in social justice at a higher level. Instead, this form of communication is purely consumer driven and serves to provide consumers' entry level into activism that requires only purchase rather than more organised types of mobilisation and campaigning. Lekakis' (2013) analysis draws on Beck's (2002, 2004) notion of 'banal cosmopolitanism' to highlight the illusion of global connection encouraged by fair trade communication, which occurs through the act of consumption of globally produced and imported goods and leads to an oversimplification of global life. This oversimplification is also underpinned by a sense of 'cosmopolitan sympathy' and sentiment towards the other, as opposed to a more reflexive and engaged expression of solidarity that should characterise citizenship. It is therefore unsurprising that consumers remain alienated and disconnected from producers (Fridell 2007; Lekakis 2013; Littler 2009).

The problem with sympathy, as an expression of activism, is that it gradually leads to the decline of public interest and engagement with sustainable development. On the one hand, consumers are increasingly becoming suspicious of fair trade's communication strategies, especially since the mainstream retailers and supermarkets became involved. One of the participants interviewed for this book commented:

> I'm a little bit cynical about the supermarkets using these stories of farmers in Africa or India to sell products, they think that psychologically we might be tapping into it because there's a picture of a guy standing next to his coffee plantation … but how much of the money that Tesco are making selling this product is he actually going to get at the end? Or he could be just an actor. (Participant, E, Group Interview 10)

On the other hand, even if fair trade's communication strategies may encourage purchase, they may harm other forms of citizen engagement, such as cognitive and behavioural engagement in the form of awareness, knowledge, or campaigning. Hussey and Curnow (2013: 52) point to the 'collective blindness' of consumers who tend to understand the world—people, places, and processes involved in food production—through a hegemonic prism, even in the absence of a poverty or sympathy framed information and visuals. The reactions of some of the consumers who were interviewed were contradictory but also quite telling in terms of their struggle to distance themselves from the role of the 'middle-class saviour'. Although participants across the different groups noticed and commented on the fact that the Wayanadan farmers looked significantly better-off than the common fair trade images portray, some of the participants still perceived the farmers as in need of 'our support'. The most telling example is the statement of one of the participants who said:

> I think we're a kind hearted race, we give a lot to charity and I'm not saying they're [the Wayanadan farmers] a charitable case but I'm just using that analogy because you see it that way—you feel you've done a little something because you're one of the lucky people in the world, you've got money and so on … I'd like to reward them for being such humble, great people, who are not making a lot of money and probably don't want to but I'd be more likely to buy their jar of honey than one made in Yorkshire … (Participant B, Group Interview 4)

At the heart of these perceptions lies the image of the post-colonial citizen of the Global South as simple and benevolent, while their intellectuality is denied (Varul 2008). This was again reflected in the comments of some of the focus group participants who described the Wayanadan farmers as 'humble', 'genuine', 'decent' and 'hard-working', despite their recognition that they also looked 'like us' (i.e. not poor). The danger here is that the legitimacy of the fair trade farmer is shaped by this stereotypical and one-dimensional image. Consumers end up understanding certain areas of the world as uniformly marginalised and dependent upon the actions of Global North businesses and consumers, which inhibits their ability to engage with distant communities in a meaningful way. In this case, intervention through consumption reinforces difference and dependence instead of solidarity and support for the agency of farming and artisan communities (Hussey and Curnow 2013; Wright 2004). The case of fair trade presents a typical example of a specific identity which consumers are encouraged to embrace. It is the identity of the benevolent saviour that is constructed through communication strategies, and which is expected to occur mainly through the act of purchase, but is less successful in encouraging the necessary consumer reflexivity and recognition of the value of mutual support, interdependence, and sense of community food environments.

Consumer Mobilisation and the Role of Information in Organic Networks

The organic food movement comprises local and global networks, so unlike in the case of fair trade, consumers' action and citizenship is not measured only in the global arena. However, with the rapid growth of the global organic market over the last decades (see Chap. 2), a considerable volume of organic products consumed in the countries of the Global North are most likely imported through the South-North organic trade. In the realm of the organic food networks, the mobilisation of consumer citizenship remains something of a conundrum, where consumers' attitudes are driven by a mixture of self-interest and concerns over the environment, while they are also faced with ethical dilemmas over the environmental implications of imported organic produce compared to local and seasonal grown food. While in the case of fair trade, consumers' actions are unequivocally aimed at supporting the distant producer, participation in the organic movement is often riddled by the local versus organic dilemma. On the one hand, such dilemmas might result in unreflexive localisation. On the other, they may encourage more careful consideration of one's consumer behaviour, carrying the potential for transformation and adoption of more sustainable lifestyles.

The consumption of organic products is perceived as an expression of sustainable citizenship and active environmentalism for it embraces production methods that are in harmony with the environment, enhance biodiversity, and are better for wildlife (DEFRA, UK Department of Environment, Food and Rural Affairs 2002). Yet, although it has been demonstrated that environmental concerns influence consumer behaviour in a favourable manner, the driving factor in different parts of the world appears to be perceptions of good health and taste (Magnusson et al. 2003; Pham et al. 2019). Consumers tend to associate organic food with personal well-being, health, and moral values, a view that was also reinforced by a project that studied consumer behaviour across different European countries (Cicia et al. 2002; Cottingham and Winkler 2007). Health was also the primary motivation for purchasing organic products among the consumers who were interviewed for this book, with several of the participants commenting on becoming more health conscious and more aware about organic products as they grew older. The following quotes demonstrate this sentiment:

> I think about what I'm putting into my body a lot more now than I used to 10 years ago…. (Participant B, Group Interview 6)

> I tend to go more for 'organic' than 'seasonal' … I don't want to exercise and be really healthy and then shove a load of chemicals into my body because I don't know what effect it's going to have down the line, so that's why I prefer to go 'organic', I know what I'm putting in my body then. (Participant B, Group Interview 7)

> I get fruit and veg and also my eggs organic because I think I'm a bit paranoid … because you don't know what's going into your food and I think, with cancer and stuff, back in the day you knew that if you smoked you were going to get cancer, but people are getting cancer and they've never smoked; and it's like, 'What is going into our food?' There's an element of just thinking, 'If I eat organic that probably will eliminate that risk for me', because of these chemicals, they don't do you any good. (Participant E, Group Interview 8)

It was therefore unsurprising that the story of the farmer in Wayanad whose daughter died of cancer (Chap. 5) was brought up several times during the discussion, signalling consumers' perception of the link between organic food and protection from life-threatening diseases. Consumers of organic food tend to be motivated by self-interest and concerns about their health rather than by feelings of solidarity and environmental considerations, an attitude that is at least partly down to the absence of the necessary public education that will raise awareness and understanding of the wide range of benefits and will also translate into consumer commitment and spending (Cottingham and Winkler 2008).

Although education and information about the connections between personal food choices and sustainability is a crucial component in the mobilisation of consumers towards social change and sustainable food consumption (Tischner et al. 2010), in the case of organic food, consumers' informational needs remain unfulfilled. Organic products have received limited advertising and explicit promotion, hence, the mobilisation of organic consumers has relied primarily on information and promotional material at the point of retail, focusing on 'chemical free' and 'environmentally friendly' qualities. But such information remains too vague to activate consumers' civic duty as citizens (Halpin and Brueckner 2004). In fact, consumers do not always understand key attributes and the complexities of organic farming practices or the costs incurred (Barrena and Sanchez 2010).

A common information channel is through labelling initiatives designed to communicate the ethical qualities of a product, helping consumers distinguish between organic and conventional items. Strictly speaking, labelling is a market-based tool that provides assurance that a product has been certified by a recognised regulatory body. Although such initiatives are meant to assist farmers to internalise the cost of production, giving them access to foreign markets, labelling and certification schemes are also replete with economic challenges for producers (see Chap. 2). From the consumers' point of view, labelling supports producer-consumer relations by linking production practices with the place of its consumption, via the global food flows, but the evidence for its effectiveness is mixed. Several studies have demonstrated the major role played by labels—as indications of the science involved in the processing of the food—in securing consumer's trust and willingness to pay a premium price for organic products, in the absence of direct contact with farmers. The US national standard implemented in 2002 was a case in point, as it boosted consumer awareness and confidence in organic products (Sahota 2004). At the same time, consumers appear sceptical towards certification bodies and the genuineness of organic products (Canavari et al. 2002; Aarset et al. 2004). More crucially, they also often lack the knowledge and confidence to make sense of food labelling or even recognise certification labels (Padel and Foster 2005). Certification bodies are responsible for educating consumers on relevant logos and labels, but this is rarely the case. This was illustrated in the introduction of the mandatory EU logo for organic food. Not being accompanied by sufficient communication, it led to consumer lack of understanding and distrust of the logo (Janssen and Hamm 2010; Kastberg 2015). Denmark is perhaps the only European country where labelling works effectively. The Danish state-certified organic logo (the red Ø-label) has helped consumer confidence in organic products reach a high level. The Danish organic logo is known by 98% of all consumers, and 90% of consumers are confident that products thus labelled are actually organic (Wier et al. 2008).

Despite questions surrounding the effectiveness of education and communication around organic products, consumers' appetite for information about the origins and story of their food is growing. A successful example of an organic brand that used the storytelling method to introduce its cocoa producers is Green & Black's, whose attempt to personalise the product proved popular particularly with female consumers above the age of 55 (Cottingham and Winkler 2008). Telling a product story in a subtle way was also seen favourably by the consumers interviewed in London and Leicester, who likened this storytelling approach with the experience of meeting the producer at the farmers' market. One of the participants stated, 'It's interesting if they put their story on the packaging, it makes you far more engaged, it gives you that human appeal' (Participant A, Group interview 9), and another one added, '...it is like when you go to a farmers' market, one of the great things is that you get to talk to the people and they are very passionate about what they're doing, you buy into the person' (Participant B, Group Interview 9).

Consumers' reported desire to feel more connected to the place and person behind the organic product highlights the importance of food provenance and, at the same time, it is indicative of the tensions and dilemmas that consumers are faced with in relation to the organic, but imported, versus local choice (Timmins 2010). This tension was reflected in the words of one of the focus groups participants who said, 'For me "organic" is the natural way, the way it used to be 100 years ago, this is how it should be. The question I often have is, "Is it really true, can I really trust that?" because I'm not buying it from that farmer, I'm just going into a supermarket and maybe buying organic, yes, but I don't know...' (Participant D, Group Interview 5). In fact, many consumers believe that for organic food production to be beneficial, it should be combined with environmental measures such as reducing food miles and involving local or regional production and consumption (Vogl et al. 2004). It is this need for trust and 'reconnection' with the different elements of the food system (Dowler et al. 2010: 205) that underpins bigger questions and discussions around the global versus local producer-consumer relations and the role of communication in building these relations. The issue of reconnection is therefore crucial because it unravels both the tensions and the opportunities that exist for closer producer-consumer relations in global food networks, and the possible new ways for consumer contribution to sustainable development. The missing voice of the producer is a crucial component in building producer-consumer relations that are more likely to encourage reflexivity and citizenship in the global food system.

Producer-Consumer (Re)Connection Through Storytelling

Consumers not only need education, but they also appreciate the personal and emotional connection with the people who grow their food (Tischner et al. 2010). This connection can materialise through a more direct and innovative producer-consumer communication that is based on content and stories that brings the reality of food production to the consumer through the words and images of the producer. The storytelling experiment introduced in Chap. 5 is here approached from the perspective of the audience. It offers evidence of different ways in which this type of communication can prove empowering for consumers by activating a sense of collaboration and a thought process that can potentially lead to more conscious and sustainable lifestyles. The power of storytelling, in this case, lies in the people's natural inclination to relate to and respond to stories (Bruner 1986). Because stories can add a layer of learning, at the human level, storytelling can be a medium that advocates pro-environmental action and ecological citizenship through narrative rather than through instruction. Storytelling becomes an instrument for people to connect to certain political issues through the power of personal stories and narrations. In this case, issues such as ethical food production and consumption transform from abstract concepts to more personal issues that concern one's own life. Hence, stories can function as a source of ideas upon which audiences could act, and as 'a complementary form of advocacy to the more direct form that relies on facts and information, but it is one that is equally if not more powerful' (Gearty et al. 2013: 16).

Stories make it easier for audiences to absorb and remember information, compared to the more factual forms of communication, and this quickly became evident in the discussion with the consumers in the UK as there was one story in particular that caught their attention and was brought up several times in the conversation. The story concerned an advertisement by Kenco, featuring a young man who escapes from gangs by becoming a coffee grower. The advertisement was part of Kenco's

Coffee vs Gangs campaign that aimed to provide young people in Honduras with opportunities to train as coffee farmers as an escape route from gang violence. The specific advert was a typical example of how a story can transform ideas into more tangible and personal issues for the audience/consumer. As one of the participants put it, 'It [the advert] makes it official, it makes it real, it makes you think about it, it makes you aware … if the advertising wasn't there then you probably wouldn't think about it at all' (Participant B, Group Interview 1). The advert is also an example of how audiences can connect with stories on an emotional, personal, and rational way at the same time, as the following quote from another participant also demonstrates: 'I buy Kenco, because of their work in South America, with the gangs, and I relate to that because I work with a lot of the gangs in Hackney, so I understand why they're doing it' (Participant D, Group Interview 1).

The stories of the Wayanadan farmers were received enthusiastically by the focus group participants, which suggests the tone and production of these videos was perceived as an appealing format. The initial reactions focused on the scenery and the pace of the videos, with participants across all groups describing them as 'beautiful', 'calming', 'peaceful', and 'atmospheric', while they also appreciated the authenticity of the farmers as storytellers. As the discussion carried on, the responses varied according to the age of the participants, with the younger groups suggesting improvements to the website, making it more interactive and dynamic with the addition of infographics and discussion forums. Some of them commented on the repetitiveness of the storyline, as many of the videos showed similar farms and crops. It is important to mention that the main objective of the experiment was to explore audiences' reactions to content that was produced by the farmers and portrayed aspects of the production process, such as the handling of bees or the making of fertilisers. Hence, although the participants' comments on the technical side of the website were welcome, the emphasis was on their processing of the information and the images they were exposed to, and on exploring the type of producer-consumer connection that this format of digital storytelling can encourage.

At this point it is also worth noting that, in the context of this experiment, storytelling provided a platform for consumer participation in the research and interaction with the farmers. It facilitated the participation of the consumers by offering the space for them to create and express their own knowledge and, in so doing, empower themselves to affect the type of change that is appropriate to them (Cornish and Dunn 2009). In addition, through the mediation of the researcher, the consumers engaged in a virtual communication with the farmers. Here, the researcher acted as a facilitator who assisted the consumers in articulating their own understanding of the farmers' stories and creating shared meanings through their interaction with each other and with the stories. Consumers renegotiated their relationships with food and its producers and constructed new knowledge that was communicated back to the farmers through the researcher. The remaining of the chapter will focus on the complex ways in which farmer-generated stories could influence consumers' pro-environmental behaviour and advocacy.

(Re)Connection Through Online Media

The proliferation of online and traditional media plays a central role in mobilising ethical consumers, through a steady flow of information on the ethics of shopping and the promotion of causes (Lekakis 2013). Moreover, the virtualisation of food networks and the producer-consumer connectivities in online spaces is quickly becoming a 'fashionable' form of consumer-citizen activism, with empowering possibilities but also unavoidable barriers and limitations (Goodman et al. 2014). Yet, the relationship between online space and reconnection and the role of online tools in facilitating more direct relations are issues that require deeper engagement with. The video experiment presented here utilised a combination of technologies, including a mobile phone video camera and an online platform/website through which the farmers' videos were shared with the consumers. The experience of the consumers demonstrated how this online interaction allows them to extend their connection to the farms and the crops through a virtual capacity. Images of the land, the produce, and the processes such as the preparation of fertilisers was something that participants reported appreciating, as the following quotations show:

> It's nice to have a visual, I imagined something in my head but to actually see the video of them picking it [the produce] and the story, it was quite nice to watch. (Participant C, Group Interview 2)

> It was like they were in a natural forest and they kept saying, 'And this is a coffee plant' and I was thinking, 'Where!? … oh my god, it's just part of the flora and fauna!' (Participant, D Group Interview 2)

> To actually see them taking the coffee beans from the tree and put them on a plastic sheet and then having to sort out the green ones from the mouldy grey, the odd one or two, it did really bring it home. (Participant, C Group Interview 3)

> I hadn't thought about it, it was just different, I expected to see a farm and a digger and a workforce but it literally was the family and two cows, or very small scale, but they were making a living from it and providing Fair Trade, organic products. (Participant A, Group Interview 10)

> A few of them sort of depicted the product they grew with healing properties, like the turmeric one and there was a specific basil, which I found very informative and useful to know. (Participant B, Group Interview, 8)

In effect, by extending the biological into the virtual, this type of virtual connection offers consumers an alternative embodied experience of the material environment of food production at a distance. In a study of Alternative Food Network schemes in England, Bos and Owen (2016) demonstrated how online spaces can generate notions of community between producers and consumers and re-create material spaces in an online capacity. In a similar way, the consumers' experience with the Wayanadan farmers' videos shows evidence of existing possibilities for social connections to happen even in spatially extended networks leading to a renewed awareness around food production. A crucial element in these social connections is the personal connection with producers, and which can be possible even in spatially extended food networks through the virtual spaces of social interaction. Several of the focus group participants said that watching the videos enabled them to link the whole idea of food production back to real people whom they could relate to. This personal connection became possible through the storytelling format, the first person voices, and visual material that was shared through technology. These elements can have a multi-sensory impact, engage people in conversation, and create a sense of more direct communication and connection (Cornwall 2016).

The transformative capacity of these social connections lies in the broader changes they can bring to mainstream food systems (Holloway et al. 2007). This is through what Dowler et al. (2010) describe as the 'graduation effect', a situation where the purchase of food outside the 'mainstream' encourages consumers to rethink and refine other lifestyle practices (Dowler et al. 2010: 210). Of course, evidencing the 'graduation' from virtual reconnections to behavioural change, citizenship, and sustainable lifestyles would require further in-depth and long-term study (Carolan 2006; Cox et al. 2008). This should also include the exploration of a more horizontal virtual interaction where producers have the opportunity to have access to information and visual content related to the consumers' dietary habits and use of certain products. This idea was raised several times in the discussions with the farmers in Wayanad, who expressed a keen interest in knowing and understanding how their products are being consumed abroad. It was also raised in the discussions with the consumers with several participants proposing the creation of interactive sites where producers and consumers could engage in a more dialogic relation. Given consumers' reliance on media technologies—from social media to more traditional media such as magazines and the radio—for food-related information (from simple things such as recipes, to the environmental impact of certain production processes), a virtual connection that relies on interactive and participatory produced content can give rise to a different type of consumer citizenship and activism more broadly. This activism will not be expressed through ethical and political consumption, but through the formation of producer-consumer virtual communities with a shared interest in sustainable food production and consumption. Besides, with 'exchange' of knowledge emerging as the next driver of development, growth, and change (Light 2014), producers and consumers can learn from each other and play an active role in determining sustainable food production. In fact, campaigning organisations argue that there are environmental, health, socio-cultural, and economic benefits to be gained from building knowledge-exchange relationships between farmers and consumers (Countryside Agency 2002). Such relationships can encourage a collaborative communication (see also Chap. 3) through the creation of joint meanings around food production and a deeper understanding of each other's role and responsibility in the global food system. In this collaboration, the role of food growers as content producers is crucial since this is the only way for consumers to expose themselves to information that does not replicate dominant narratives. At the same time, and as discussed in Chap. 5, the farmer-generated content is crucial for consumers to recognise the role of producers as equal partners in this relationship.

Disrupting the Colonial Image of the Distant Other

Chapter 5 explored the potential impact that producers' participation in the creation of the narratives through which they are represented can have on their own development and well-being. Yet, producers' capacity to represent themselves can also prove significant for consumers' ability to relate to the producers as equal partners. The power of these narratives lies in their capacity to disrupt consumers' long-standing perceptions of the distant other, and through this, build a renewed understanding of sustainable development and the role that different actors play in this. There are several pre-conceptions that the farmers' videos disrupted, but the dominant one was undoubtedly the assumption that farming is a simple and easy manual labour work. It is especially fair trade work that tends to be portrayed as simple labour, in disregard of the skills required (Varul 2008). In contrast, the videos of the Wayanadan farmers depicted them as skilful experts with the capacity to educate, and this was verified in the consumers' reactions and recognition of the educational value of these videos.

Through the visual power of the video, the Wayanadan farmers challenged another pre-conception of the farmer being in the position of 'victim' in need of intervention and dependent on the purchasing and compassion of the consumer (Hussey and Curnow 2013; Varul 2008). By offering a glimpse of the farmers' lifestyle, the videos disrupted consumers' perception of the farmer as living in poverty, struggling and being unhappy. First, there was an almost unanimous surprise and even 'shock' at the farmers' apparently comfortable lifestyle, the beautiful big houses, and the fact that, as some of the participants said, 'They seemed quite wealthy' and 'They have windows and grills on the windows'. The fact that consumers even found similarities with their own lives and families, claiming 'They are just like us', somehow reduces the distance and challenges assumptions of 'difference' and 'inferiority' of the Southern producer, enabling consumers to relate to farmers as people rather than as victims (see also Touri 2020). Second, the farmers' videos reflected a sense of positivity that disrupted consumers' pre-conceptions of development and well-being as a quality that is mainly associated with Western lifestyles and progress. This was manifested in participants' statements about the farmers looking 'content and fulfilled', 'very positive and enjoying their work', 'proud', and 'future-thinking'.

These disruptions demonstrate how a more direct and innovative producer-consumer communication enables consumers to become 'allies' rather than 'saviours' in this relationship. This requires that consumers recognise producers' agency and ability to define their own ideas of well-being, as well as the non-universality of development altogether. In this case, consumers' exposure to alternative types of content and farmers' representations can help shift established assumptions of development as a Eurocentric notion of reform, improvement, and emulation of Western criterial of progress (Escobar 2001). It can also create space for more genuine encounters and authentic connections between distant strangers, which is a pre-requisite for a cosmopolitan citizenship that is not based on an idea of cosmopolitanism as a commodity but as an ethic (Beck 2006; Binnie and Skeggs 2004). Once again, we can observe the potential for emerging collaborative communities that are based on more egalitarian relations and shared values.

Knowledge Exchange and Consumer Education for a Sustainable Future

Although Alternative Food Networks bring opportunities for further education around the origins and production of food, consumer education remains a challenge while consumers become more demanding and critical towards food-related information (Trienekens et al. 2012). Consumers need to be re-educated to be able to resist conforming to the conventional food system (Jaffe and Gertler 2006). The need for education was one of the critical issues that was raised in the focus groups, with the majority of participants admitting there was not enough useful information around the significance of participating in movements such as fair trade or organics. As one participant explained:

> I do not think there's enough information out there about Fair Trade and 'organic'. I think people are frightened by the words! If you asked the average person on the street, 'What is 'organic'?', a lot of them would be embarrassed because they don't actually know what it means. I do not think there's enough information. What I've read today, what you've asked me to do [watch the farmers' stories], I should have done that light years ago and I don't know why I haven't!? … (Participant A, Group Interview 5)

One of the key benefits of education on food origins, agricultural practices, and their impacts is the empowerment of consumers to become citizens that are reflexive and aware of their food habits and choices (Levkoe 2006). As the focus group discussions also demonstrated, consumers feel they could have a bigger impact on the global food system if they knew more about the interconnections between the choices, habits, and practices of producers and consumers across the globe. The following quote summarises this idea:

> In many ways food and eating and choosing and buying food is an education, the whole of our lives is about education and, essentially, if you're not informed then you're not going to change; you've become comfortable with what you're doing, it's normal, whatever 'normal' is ... but then again, as a community educator, I think unless we run projects that enable people to taste/sample, get informed then they won't change. (Participant A, Group Interview 6) (see also Touri 2020).

As other studies have also demonstrated, education about issues such as the seasonality of ecological production, and the difficulties producers face, can bring changes in dietary habits, and encourage the mobilisation of consumers in campaigns, which can then inform public policies on sustainable food production and consumption practices (Darolt et al. 2016). This is a slow process that takes time, and which also makes younger people's education even more crucial. A relevant point that was raised in the discussion with the consumers in the UK concerned children's lack of basic knowledge on food. A participant, who also works in education, said:

> ...it's like taking children to a supermarket and then taking them to a farm and they don't realise that those animals are wrapped in plastic, that those are cows, or the little lamb that you're going to have for Easter, you know ... kids are the future and you have to teach them, whatever class I'm teaching I'm always going to have something with dirt, I'm always going to have something with planting and always talk about the environment and good health—'Healthy eating' should be a huge part of the curriculum actually .... (Participant E, Group Interview 4)

The role of educational institutions as well as social movements is crucial for stimulating citizens to become active through school curricula, awareness campaigns, and political lobbying (Dubuisson-Quellier et al. 2011). There is also need for more innovative approaches to education that would encourage more dialogic and interactive methods. These should involve tighter communication links and feedback loops between consumers and producers, and knowledge about the farms where food is produced (Marsden et al. 2000; Sundkvist et al. 2005). It is through these links that consumers can make associations with the place and method of production and the people involved (Darolt et al. 2016). Although it might not be possible to replicate the face-to-face connections in spatially extended networks, innovative communication methods can provide opportunities for shortening the geographical distance. As Lamine et al. (2012) state, connecting producers and consumers beyond geographical distance may be necessary, especially for food that is not produced locally. This is important for consumers' understanding and appreciation of the distant producers' resilience and ability to preserve farming options and traditional methods (Clancy and Ruhf 2010). The video experiment demonstrated how the visual images of the farms and the information about production methods proved particularly educational for consumers. The following quotes reflect this:

> I didn't realise how dependent upon pesticides [farming is], and I thought they were very important in farming, I didn't realise just how many organic alternatives there were, so I've been educated and, as I say, a half hour programme, everybody who will have watched it in Britain tonight would be educated in some small way, definitely, I'm sure of it, I'm sure of it. (Participant D, Group Interview 4)

> I was interested in the little stories about their lives and I found out so many new things, or understanding the bigger picture, for me it was interesting. (Participant E, Group Interview 4)

> I think I've gained something from it, so I'd happily point someone in the direction of it, 'Have a look at that'. If someone asked me why I have organic food or something like that, it would be easier to explain that to them than to explain why I do it. (Participant D, Group Interview 9) (see also Touri 2020).

The educational value of the videos and the consumers' interest in the production process became obvious also by the fact that the focus group participants conducted their own online research about certain farming methods that were unfamiliar with, such as 'zero budget' farming. In addition, they showed a keen interest in discovering more about the entire cycle of the farming process through the four seasons, the health benefits of certain products such as turmeric, the effectiveness of organic fertilisers, and how many workers are involved in the cultivation of these farms. Many consumers also reported they were curious to know who the main customers and markets were for the farmers' products, and whether they felt that organic agriculture would produce enough food to feed future generations.

Three key observations can be drawn at this point. First, consumers' appreciation of the educational quality of the videos demonstrated a general interest in food production processes that was independent of the proximity in location. In a similar way in which knowledge transfer happens in localised food systems (e.g. CSA schemes) through the virtual spaces of online and social media networks (see Zapico and Söderberg 2018), it can take place in spatially extended food networks with the contribution of food growers from across the world, especially export countries. This can facilitate a global producer-consumer network, giving consumers access to valuable information about the material reality of food production across the globe, a reality that is not narrowed to the standard, stereotypical narratives promoted through fair trade campaigns but extends to practical information about health benefits and environmental impacts. Second, only few of the focus group participants suggested that the videos would be more valuable if they were linked to specific products that could be purchased. One participant said: 'I think it would be more interesting for me if I'd gone out and bought it and then seen all this, so that's just down to circumstances' (Participant C, Group Interview 7). On the one hand, this shows consumers' need to make sense of this information in relation to their own purchasing habits. On the other, the fact that the overwhelming majority showed a genuine interest in knowing more about these farmers' work indicates there is an appetite for building closer relations with producers based on a shared interest in the origins and production of food. The personal connection enabled by the storytelling approach played a key role in this case. Not only does it help consumers relate to food production by adding a layer of learning at the human level, it also generates further interest in knowing more about the farmers' work and life. As consumers across all the groups attested, it was this visual and personal interaction that increased their appetite for more information. The third and last observation concerns the shift from a consumer to a citizen mind-set. The consumers' reactions to and interest in the farmers' work is an indication of the leverage that closer producer-consumer relations can provide for the creation of better informed and active citizens. In this instance, we can see the potential for emerging global communities of practice that are based on processes of learning through participation in a virtual interaction between producers and consumers; and are bound by shared interests in food, even if they have weaker ties. It is also an indication of the need for governmental and non-governmental organisations and for businesses to work closely with both producers and consumers and support the creation of those communication channels and knowledge-exchange mechanisms that will nourish our relationship with food in ways that benefit the environment and the people.

Conclusion

The role of consumers in the global food market is a highly contested issue and the contribution that consumers can make to sustainable development through their purchase and lifestyle choices remains debatable. What is more certain is the increasing need of consumers to feel more connected to the origins of the food they eat. Driven by health-related food insecurities and environmental concerns, consumers in the developed economies are becoming prone to adopting defensive and unreflexive attitudes in favour of locally sourced food. Yet, such positions imply a restricted global and cosmopolitan citizen role in the promotion of sustainability.

In this chapter, the role of consumers was narrowed down to the fair trade and organic movements as examples of globalised Alternative Food Networks where producer-consumer connections occur through the respective labelling and information schemes. Although the type of knowledge disseminated in each case is different—while organic food labelling focuses on aspects of the production process, fair trade material aims to invoke empathy and sympathy towards the distant farmer—in both cases this knowledge fails to adequately educate consumers and inform a reflexive attitude towards food and the people who produce it. Moreover, it often reinforces consumer fantasies and imaginaries about what fair trade and organic food should be about, which producers are expected to fulfil. Drawing on consumers' reported need for trust and 'reconnection' with the different elements of the food system, this chapter employed the notion and practice of 'weak' communication in the form of storytelling and digital connection between consumers in the North and producers in the South that was materialised through the less formal farmers' own videos, images, and written stories. The storytelling exercise that was introduced in Chap. 5 was here approached with a focus on the capacity of a more personalised and participatory communication to inform a lifestyle politics where consumers are encouraged to reflect on their everyday lives and choices, and the narratives attached to those (Spaargaren and Oosterveer 2010).

The main lesson to be gained from this experiment can be summarised in the power of the personal approach of the farmers' videos to perform a pedagogical function as well as elicit what seemed like a genuine consumer interest in understanding aspects of the food production process. The need for consumers to be addressed in an emotional way is a recognised fact (Tischner et al. 2010: 276–277), but the farmers' videos managed to achieve this emotional connection while at the same time conveying practical and rational information concerning their products and practices. This combination activated a thought process where consumers were able to question and reflect on some of their established pre-conceptions about the organic food production, the image of the fair trade farmers and the relationality of the local and global. Supported by the advantages of digital technologies, the personal connection enabled by the farmers' videos made up for the geographical distance between producers and consumers in a globalised setting.

In the case of consumers' contribution to the sustainability of global food networks as agents of change, the findings highlight the crucial role that a more interpersonal communication with the distant producer can play in enabling them to reconsider their existing perceptions and behaviours. An example of such reconsideration can be found in consumers' redefined perception of the distant farmer as a knowledgeable professional with the capacity to educate. Such perceptions can encourage and pave the way for a renewed sense of producer-consumer solidarity that is based on the creation of joint meanings around food production and a shared interest for sustainable food production and consumption; and this can find expression in action that goes beyond just the act of purchase, such as mobilisation and campaigning. In this case, we see the emergence of a discursive and communicative collaboration between producers and consumers that is laden with shared values on issues related to food, health, and the environment. The empowering role of a more direct and horizontal communication, such as the Wayanadan farmers' videos, lies in its emotive and embodied nature that can provide a thick description of food production and can convey complex information putting consumers into a scene and inviting them to challenge their viewpoints. This kind of communication can also 'educate' consumers in making sense of the complex and competing values of their role as global cosmopolitan citizens.

The power of weak communication is here embodied in the more informal and personal connection that became possible through the storytelling experiment, while its contribution to sustainable development can happen through the new meanings and understandings of food production and consumption that consumers draw from this communication, and the impact that such meanings can have on their perceptions and behaviours. Weak communication is also manifested through a reconceptualisation of the role of communication in sustainable development, which in this instance happens through the producer-consumer interaction in the context of the global food economy. Although such interactions may be difficult to replicate or make scalable, the insights drawn from this experiment can pave the way for new approaches to communication and education around issues of food production and consumption.

Institutions, organisations, and cultural intermediaries with an appetite to facilitate sustainable food education and connections between personal behaviour and food choices, well-being, and sustainability can play a key role. Non-governmental organisations play a strategic role in orienting consumer behaviour by publicising the connections between consumer choices and action and the implications for sustainability (Dubuisson-Quellier and Lamine 2008). They could also facilitate a more horizontal communication between producers and consumers. The same applies to retailers, such as the SMEs presented in Chap. 4. Many of these companies have a genuine social and economic interest in supporting the farmers. Investing in connecting their consumers with food growers and suppliers could strengthen the companies' socially and environmentally responsible image, sales potential, as well as foster goodwill and working relationships with the farmers. What can be said with certainty is that the genuinely sustainable products have a good story to tell, and it is this story that can give the impetus for organisations and businesses to play a pivotal role in building more meaningful and creative producer-consumer relations. It is through such relations that the value of communication for sustainable development can be further recognised and appreciated.

References

Aarset, B., Beckmann, S., Bigne, E., Beveridge, M., Bjorndal, T., Bunting, J., McDonagh, P., Mariojouls, C., Muir, J., Prothero, A., Reisch, L., Smith, A., Tveteras, R., & Young, J. (2004). The European consumers' understanding and perceptions of the "organic" food regime: The case of aquaculture. *British Food Journal, 106*(2), 93–105. https://doi.org/10.1108/00070700410516784.

Amin, A. (2002). Spatialities of globalisation. *Environment and Planning A, 34*(3), 385–400.

Barnett, C., Clarke, N., Cloke, P., & Malpass, A. (2005). The political ethics of consumerism. *Consumer Policy Review, 15*(2), 45–51.

Barrena, R., & Sanchez, M. (2010). Frequency of consumption and changing determinants of purchase decision: From attributes to values in the organic food market. *Spanish Journal of Agricultural Research, 8*(2), 251–272.

Bauman, Z. (2007). *Liquid times: Living in an age of uncertainty.* Cambridge: Polity.

Beck, U. (2002). Cosmopolitan society and its enemies. *Theory, Culture and Society, 19*(1–2), 17–44.

Beck, U. (2004). Cosmopolitan realism: On the distinction between cosmopolitanism in philosophy and the social sciences. *Global Networks, 4*(2), 131–156.

Beck, U. (2006). *The cosmopolitan vision.* Cambridge: Polity.

Bennett, W. L. (2003). New media power: The internet and global activism. In N. Couldry & J. Curran (Eds.), *Contesting media power: Alternative media in a networked world* (pp. 17–37). Oxford: Rowman and Littlefield.

Bennett, W. L., & Segerberg, A. (2012). 'The logic of connective action, 'information. *Communications Society, 15*(1), 739–768.

Binnie, J., & Skeggs, B. (2004). Cosmopolitan knowledge and the production and consumption of sexualized space: Manchester's Gay Village. *The Sociological Review, 52*(1), 39–61. https://doi.org/10.1111/j.1467-954X.2004.00441.x.

Bos, E., & Owen, L. (2016). Virtual reconnection: The online spaces of alternative food networks in England. *Journal of Rural Studies, 45*(1), 1–14.

Boström, M., Micheletti, M., & Oosterveer, P. (2019). Studying political consumerism. In M. Boström, M. Micheletti, & P. Oosterveer (Eds.), *The Oxford handbook of political consumerism* (pp. 1–25). Oxford: Oxford University Press.

Booth, S., & Coveney, J. (2015). *Food democracy. From consumer to food citizen.* Singapore: Springer.

Bruner, J. (1986). *Actual minds, possible worlds.* Cambridge, MA: Harvard University Press.

Canavari, M., Bazzani, G. M., Regazzi, D., & Spadoni, R. (2002). Food safety and organic fruit demand in Italy: A survey. *British Food Journal, 104*(3/4/5), 220–232. https://doi.org/10.1108/00070700210425688.

Carolan, M. S. (2006). Do you see what I see? Examining the epistemic barriers to sustainable agriculture. *Rural Sociology, 71*(2), 232–260. https://doi.org/10.1526/003601106777789756.

Castree, N. (2004). Differential geographies: Place, indigenous rights and 'local' resources. *Political Geography, 23*, 133–167.

Childs, J. B. (2003). *Transcommunality: From the politics of conversion to the ethics of respect*. Philadelphia: Temple University Press.

Cicia, G., Del Giudice, T., & Scarpa, R. (2002). Consumers' perception of quality in organic food: A random utility model under preference heterogeneity and choice correlation from rank-orderings. *British Food Journal, 104*(3/4/5), 200–213. https://doi.org/10.1108/00070700210425660.

Canclini, N. G. (2001). *Consumers and citizens: Globalization and multicultural conflicts*. Minneapolis: University of Minnesota Press.

Clancy, K., & Ruhf, K. (2010). Is local enough? Some arguments for regional food systems. *Choices: The Magazine of Food, Farm, and Resource Issues, 25*(1), 1–5.

Cornish, L., & Dunn, A. (2009). Creating knowledge for action: The case for participatory communication in research. *Development in Practice, 19*(4–5), 665–677. https://doi.org/10.1080/09614520902866330.

Cornwall, A. (2016). Women's empowerment: What works? *Journal of International Development, 28*(3), 342–335. https://doi.org/10.1002/jid.3210.

Cottingham, M., & Winkler, E. (2007). The organic consumer. In S. Wright & D. McCrea (Eds.), *The handbook of organic and fair trade food marketing* (pp. 29–53). Oxford: Blackwell Publishing.

Cottingham, M., & Winkler, E. (2008). The organic consumer. In S. Wright & D. Mccrea (Eds.), *The handbook of organic and fair trade food marketing* (pp. 29–53). New York, NY: Wiley.

Countryside Agency. (2002). *Countryside agency policy on rural social exclusion 'Rural Social Exclusion AP01/54'*. Available at: www.countryside.gov.uk

Cox, R., Holloway, L., Venn, L., Dowler, L., Hein, J. R., Kneasfey, M., & Tuomainen, H. (2008). Common ground? Motivations for participation in community-supported agriculture scheme. *Local Environment, 13*(3), 203–218. https://doi.org/10.1080/13549830701669153.

Dalton, R. J. (2008). Citizenship norms and the expansion of political participation. *Political Studies, 56*(1), 76–98. https://doi.org/10.1111/j.1467-9248.2007.00718.x.

Darolt, M. R., Lamine, C., Brandenburg, A., FaggionAlencar, M. D. C., & Abreu, L. S. (2016). Alternative food networks and new producer-consumer relations in France and in Brazil. *Ambiente & Sociedade, São Paulo, 19*(2), 1–22.

Defra (UK Department of Environment, Food and Rural Affairs). (2002). *The strategy for sustainable farming and food: Facing the future*. London: Defra.

Dowler, E., Kneafsey, M., & Cox, R. (2010). "Doing food differently": Reconnecting biological and social relationships through care for food. *The Sociological Review, 57*(2), 200–221. https://doi.org/10.1111/j.1467-954X.2010.01893.x.

Dubuisson-Quellier, S., & Lamine, C. (2008). Consumer involvement in fair trade and local food systems: Delegation and empowerment. *GeoJournal, 73*, 55–65.

Dubuisson-Quellier, S., Lamine, C., & Le Velly, R. (2011). Citizenship and consumption: Mobilisation in alternative food systems in France. *Sociologia Ruralis, 51*(3), 304–323. https://doi.org/10.1111/j.1467-9523.2011.00540.x.

DuPuis, E. M., & Goodman, D. (2005). Should we go "home" to eat?: Toward a reflexive politics of localism. *Journal of Rural Studies, 21*(3), 359–371.

Eden, S., Bear, C., & Walker, G. (2008). Mucky carrots and other proxies: Problematizing the knowledge-fix for sustainable and ethical consumption. *Geoforum, 39*, 1044–1057.

Evans, D., & Jackson, T. (2008). Sustainable consumption: Perspectives from social and cultural theory. RESOLVE working paper 05–08. Available at: https://resolve.sustainablelifestyles.ac.uk/sites/default/files/RESOLVE_WP_05-08.pdf

Feagan, R. (2007). The place of food: Mapping out the 'local' in local food systems. *Progress in Human Geography, 31*(1), 23–42.

Fonte, M. (2013). Food consumption as social practice: Solidarity purchasing groups in Rome, Italy. *Journal of Rural Studies, 32*, 230–239. https://doi.org/10.1016/j.jrurstud.2013.07.003.

Food Ethics Council. (2017). Food citizenship: How thinking of ourselves differently can change the future of our food system findings of the future of food: A collaborative innovation project. Available at https://www.foodethicscouncil.org/app/uploads/FoodCitizenship2017_web.pdf. Accessed 15 July 2019.

Fridell, G. (2007). *Fair trade coffee: The prospects and pitfalls of market-driven social justice*. Toronto: University of Toronto Press.

Gasparėnienė, L., Remeikienė, R., & Schneider, F. G. (2017). Concept, motives and channels of digital shadow economy: Consumers' attitude. *Journal of Business Economics and Management, 18*(2), 273–287. https://doi.org/10.3846/16111699.2016.1214620.

Gearty, M., Williams, M., Pivcevic, P., & Reason, P. (2013). *Piloting digital storytelling and action research as an approach to stimulate pro-environmental advocacy and behaviour change: A report to the department for environment, food and rural affairs*. London: University of Bath, Defra.

Goodman, D., DuPuis, M. E., & Goodman, M. K. (2014). *Alternative food networks: Knowledge, practice and politics*. New York: Routledge.

Halpin, D., & Brueckner, M. (2004). *The retail pricing, labelling and promotion of organic food in Australia. The Australian organic industry: A profile*. Canberra: Department of Agriculture, Fisheries, and Forestry.

Harvey, D. (2001). *Spaces of hope*. Berkeley: University of California Press.

Hendrickson, M., & Heffernan, W. (2002). Opening spaces through relocalization: Locating potential resistance in the weaknesses of the global food system. *Sociologia Ruralis, 42*(4), 347–369. https://doi.org/10.1111/1467-9523.00221.

Hinrichs, C. (2003). The practice and politics of food system localization. *Journal of Rural Studies, 19*(1), 33–45.

Hinton, E., & Goodman, M. (2010). Sustainable consumption: Developments, considerations and new directions. In M. Redclift & G. Woodgate (Eds.), *The international handbook of environmental sociology* (pp. 245–261). Cheltenham: Edward Elgar.

Holloway, L., Kneafsey, M., Cox, R., Venn, L., Dowler, E., & Tuomainen, H. (2007). Beyond the 'alternative'-'conventional' divide?: Thinking differently about food production-consumption relationships. In D. Maye, L. Holloway, & M. Kneafsey (Eds.), *Alternative food geographies: Representation and practice* (pp. 77–93). Oxford: Elsevier.

Hussey, I., & Curnow, J. (2013). Fair trade, neocolonial developmentalism, and racialized power relations. *Interface: A Journal for and About Social Movements, 5*(1), 40–68.

Jackson, T. (2006). Consuming paradise? Towards a social and cultural psychology of sustainable consumption. In T. Jackson (Ed.), *The Earthscan reader in sustainable consumption* (pp. 365–397). London: Earthscan.

Jaffe, J., & Gertler, M. (2006). Victual vicissitudes: Consumer deskilling and the (gendered) transformation of food systems. *Agriculture and Human Values, 23*, 143–162. https://doi.org/10.1007/s10460-005-6098-1.

Janssen, M., & Hamm, U. (2010). Standards und Kennzeichen für Öko-Lebensmittel aus Verbrauchersicht: Empfehlungen für agrar-politische Entscheidungsträger. *Berichte über Landwirtschaft, 88*(1), 86–10.

Jarosz, L. (2008). The city in the country: Growing alternative food networks in metropolitan areas. *Journal of Rural Studies, 24*(3), 231–244.

Kastberg, P. (2015). Promoting communication, participation, and learning with regard to organic food products: A communication theoretical approach. *Ecology and Society, 20*(1). https://doi.org/10.5751/ES-07139-200103.

Klein, N. (2010). *No space, no choice, no jobs, no logo*. London: Fourth Estate.

Kumpulainen, K., & Kaartinen, S. (2003). The interpersonal dynamics of collaborative reasoning in peer interactive dyads. *The Journal of Experimental Education, 71*(4), 333–370. https://doi.org/10.1080/00220970309602069.

Kurian, P., Munshi, D., & Bartlett, R. V. (2014). Sustainable citizenship for a techno-logical world: Negotiating deliberative dialectics. *Citizenship Studies, 18*(3–4), 393–409.

Lamine, C., Renting, H., Rossi, A., Wiskerke, J. S. C., & Brunori, G. (2012). Agri-food systems and territorial development: Innovations, new dynamics and changing governance mechanisms. In I. Darnhofer (Ed.), *Farming systems research into the 21st century: The new dynamic* (pp. 229–256). Dordrecht: Springer - ISBN 9789400745025.

Lekakis, E. (2013). *Coffee activism and the politics of fair trade and ethical consumption in the global north: Political consumerism and cultural citizenship*. Basingstoke: Palgrave Macmillan.

Levkoe, C. Z. (2006). Learning democracy through food justice movements. *Agriculture and Human Values, 23*(1), 89–98.

Light, A. (2014). The allure of provenance: Tracing food through user-generated production information. In J. H. J. Choi, M. Foth, G. Hearn, J. S. Baek, P. Barden, E. P. S. Baumer, E. Blevis, R. Comber, J. Duruz, & K. Frosch (Eds.), *Eat, cook, grow: Mixing human-computer interactions with human-food interactions* (pp. 213–226). Cambridge: MIT Press.

Littler, J. (2009). *Radical consumption: Shopping for change in contemporary culture*. Berkshire: Open University.

Lockie, S. (2009). Responsibility and agency within alternative food networks: Assembling the "citizen consumer". *Agriculture and Human Values, 26*(3), 193–201.

Magnusson, M. K., Arvola, A., Koivisto Hurstia, U. K., Åbergb, L., & Sjödéna, P. O. (2003). Choice of organic foods is related to perceived consequences for human health and to environmentally friendly behaviour. *Appetite, 40*(2), 109–117. https://doi.org/10.1016/S0195-6663(03)00002-3.

Marsden, T., Banks, J., & Bristow, G. (2000). Food supply chain approaches: Exploring their role in rural development. *Sociological Ruralis, 40*(4), 424–438.

Massey, D. B. (1994). *Space, place, and gender*. Minneapolis: University of Minnesota Press.

McGregor, S. L. T (2002, April). *Consumer citizenship: A pathway to sustainable development? Keynote at international conference on developing consumer citizenship*. Hamar: Norway.

Micheletti, M. (2003). *Political virtue and shopping: Individuals, consumerism, and collective action*. New York: Palgrave Macmillan.

Micheletti, M., & Stolle, D. (2012). Sustainable citizenship and the new politics of consumption. *The annals of the American Academy of Political and Social Science., 644*(1), 88–120. https://doi.org/10.1177/0002716212454836.

Opitz, I., Specht, K., Piorr, A., Siebert, R., & Zasada, I. (2017). Effects of consumer-producer interactions in alternative food networks on consumers' learning about food and agriculture. *Moravian Geographical Reports, 25*(3), 181–191. https://doi.org/10.1515/mgr-2017-0016.

Padel, S., & Foster, C. (2005). Exploring the gap between attitudes and behaviour: Understanding why consumers buy or do not buy organic food. *British Food Journal, 107*(8), 606–625. https://doi.org/10.1108/00070700510611002.

Pham, T. H., Nguyen, T. N., Phan, T. T. H., & Nguyen, T. (2019). Evaluating the purchase behaviour of organic food by young consumers in an emerging market economy. *Journal of Strategic Marketing, 27*(6), 540–556. https://doi.org/10.1080/0965254X.2018.1447984.

Reisch, L. A. (2004). Principles and visions of a new consumer policy. *Journal of Consumer Policy, 27*, 1–27.

Renting, H., Marsden, T., & Banks, J. (2003). Understanding alternative food networks: Exploring the role of short food supply chains in rural development. *Environment and Planning A, 35*, 393–411.

Ritzer, G. (2003). *The McDonaldization of society, new* (Century ed.). Thousand Oaks: Pine Forge Press.

Sage, C. (2014). The transition movement and food sovereignty: From local resilience to global engagement in food system transformation. *Journal of Consumer Culture, 14*(2), 254–275. https://doi.org/10.1177/1469540514526281.

Sahota, A. (2004). Changing retail landscape for organic products. *The European Regional Review, Winter, 44*, 50–53.

Sassatelli, R. (2007). *Consumer culture: History, theory and politics*. London: SAGE.

Seyfang, G. (2006). Ecological citizenship and sustainable consumption: Examining local food networks. *Journal of Rural Studies, 22*(4), 385–395.

Smith Maguire, J. (2012). Provenance as a filtering and framing device in the qualification of wine. *Consumption Markets & Culture, 16*(4), 368–391. https://doi.org/10.1080/10253866.2012.662829.

Smith, M. J., & Pangsapa, P. (2008). *Environment and citizenship: Integrating justice, responsibility and civic engagement*. London: Zed Books.

Smithers, J., Lamarche, J., & Joseph, A. E. (2008). Unpacking the terms of engagement with local food at the farmers' market: Insights from Ontario. *Journal of Rural Studies, 24*(3), 337–350. https://doi.org/10.1016/j.jrurstud.2007.12.009.

Soper, K. (2007). Re-thinking the "good life": The citizenship dimension of consumer disaffection with consumerism. *Journal of Consumer Culture, 7*(2), 205–229.

Spaargaren, G., & Oosterveer, P. (2010). Citizen-consumers as agents of change in globalizing modernity: The case of sustainable consumption. *Sustainability, 2*, 1887–1908.

Stolle, D., Hooghe, M., & Micheletti, M. (2005). Politics in the supermarket: Political consumerism as a form of political participation. *International Political Science Review, 26*(3), 245–269.

Sundkvist, Å., Milestad, R., & Jansson, A. M. (2005). On the importance of tightening feedback loops for sustainable development of food systems. *Food Policy, 30*, 224–239.

Timmins, C. (2010). *Consumer attitudes towards organic food, report prepared for Organic Centre Wales.* Available at http://www.organiccentrewales.org.uk/uploads/ca_survey_br_phase_2_report.pdf. Accessed 22 Aug 2015.

Tischner, U., Stø, E., & Tukker, A. (2010). Sustainable consumption and production (SCP) of food: Overall conclusions on SCP in the food and agriculture domain. In U. Tischner, E. Stø, U. Kjærnes, & A. Tukker (Eds.), *System innovation for sustainability 3: Case studies in sustainable consumption and production – Food and agriculture* (pp. 227–285). Greenleaf Publishing Ltd: Sheffield.

Touri, M. (2020). Representation, voice and producer–consumer communication in fair trade movements: Toward new empirical directions, Geoforum, 117:103–113 doi.org/https://doi.org/10.1016/j.geoforum.2020.09.011

Trentmann, F. (2007). Citizenship and consumption. *Journal of Consumer Culture, 7*(2), 147–158.

Trienekens, J. H., Wognum, P. M., Beulens, A. J. M., & van der Vorst, J. G. A. J. (2012). Transparency in complex dynamic food supply chains. *Advanced Engineering Informatics, 26*(1), 55–65.

Van Ham, P. (2010). *Social power in international politics.* New York/London: Routledge.

Varul, M. Z. (2008). Consuming the Campesino. *Cultural Studies, 22*(5), 654–679.

Vogl, C. R., Axmann, P., & Vogl-Lukasser, B. (2004). Urban organic farming in Austria with the concept of Selbsternte ('self-harvest'): An agronomic and socio-economic analysis. *Renewable Agriculture and Food Systems, 19*(2), 67–79. https://doi.org/10.1079/RAFS200062.

Whatmore, S., & Thorne, L. (1997). Nourishing networks: Alternative geographies of food. In D. Goodman & M. Watts (Eds.), *Globalising food: Agrarian questions and global restructuring.* London: Routledge.

Wheeler, K. (2012). *Fair trade and the citizen-consumer: Shopping for justice?* Basingstoke: Palgrave.

Whitehead, M., & Bullen, A. (2005). Negotiating the networks of space, time and substance: A geographical perspective on the sustainable citizen. *Citizenship Studies, 9*(5), 499–516. https://doi.org/10.1080/13621020500301270.

Wier, M., O'Doherty Jensen, K., Andersen, L. M., & Millock, K. (2008). The character of demand in mature organic food markets: Great britain and Denmark compared. *Food Policy, 33*(5), 406–421.

Wright, C. (2004). Consuming lives, consuming landscapes. *Journal of International Development, 16*(5), 665–680.

Zapico, J. L., & Söderberg, M. (2018). *Transparent farmers: How farmers are using technology for new ways of selling and communicating with consumers.* Paper presented at the 5th international conference on information and communication technology for sustainability. Available at: https://easychair.org/publications/paper/DWMK

CHAPTER 7

Conclusion: Building Bridges for a Sustainable Future

In Wayanad, farmers are committed to protecting the unique biodiversity and the health of the current and future generations, but their efforts are met with mounting pressures from an increasingly consumerist and materialistic local society that ends up 'downgrading' the social status of organic farmers. In almost every conversation I had with the farmers, the issue of 'ridicule' they faced from local society was brought up. Equally, the recognition they received from foreign buyers and customers and the interest that their videos have generated among non-organic farmers about their work were greeted with enthusiasm and motivation to work harder and ensure younger generations had a reason to stay in Wayanad and continue farming organically.

The experience of these farmers reflects the complex and integrated nature of sustainable development, where the environmental, social, and economic dimensions are impossible to separate. It is also a reflection of the values that matter to communities and that underlie key concepts in sustainability, including vulnerability, resilience, responsibility towards future generations, and the different ways in which different communities define their problems and solutions (Miller et al. 2014). Understanding these values and the integrated nature of sustainability requires a co-creation of knowledge through collaboration between disciplines and actors to achieve the societal transformations required for the materialisation of sustainable development (Trencher et al. 2013). But until now, even if the three pillars of sustainable development—economic,

M. Touri, *Sustainable Development and Communication in Global Food Networks*, https://doi.org/10.1007/978-3-030-46119-5_7

environmental, and societal—have been the focus of scientific and policy debate, the extent to which the respective disciplines of economics, sociology, and ecology have also integrated their definitions, approaches, and findings remains debatable (Ciegis et al. 2009; Dimitrov 2010; Gibson 2006). Communication can offer a valuable conceptual and empirical tool for a more integrated understanding of sustainable development, and Servaes and Lie (2013) point to the wealth of sustainable social change work that is being done in various fields, including agriculture, food security, and natural resource management, and which involves communication. Yet, such work is rarely, if at all, framed under the heading of communication studies.

Through its focus on communication and development in the context of food networks, this book has made a step towards demonstrating the importance of relations and communication between key actors in the spheres of production and consumption and in the global/macro and local/micro levels of the food economy. These relations bring to light the different, and often subtle, ways in which the actors' interdependence can address sustainability and social change challenges. More importantly, the focus on communication and social relations enables a closer focus on people and their unique, traditional, and non-expert knowledge that can prove critical for wider social learning and capacity building. These relations and interdependencies underline the need for transdisciplinary, innovative, and participatory research (Carpiano 2009). In fact, the need for new partnerships is written into the Sustainable Development Goals as an essential path to their implementation (Trott et al. 2018). In this effort, a focus on communication is critical for analysing the links between food, place, and people as essential dimensions to sustainable development (Vandecandelaere et al. 2009). In this book, the integral role of communication in global food networks is seen in at least three areas: trade relations between food producers and small businesses, self-representation in public narratives and discourses, and digital communication between food producers and consumers. From these areas, I would highlight three emerging avenues for further research and innovation: (a) the need for small-scale projects; (b) the role of participatory and non-broadcast communication; and (c) the integration of ideas, actors, and practices from the private sector and the world of business in the field of communication and social change.

Sustainable Development: Small, Beautiful, Local, and Connected

In 1973 German-born British economist E. F. Schumacher published a book that would popularise critiques to Western neo-classical economics and would challenge the twentieth century's fixation with the idea that 'bigger is better'. The book's title, *Small is Beautiful*, captured the essence of empowerment that comes from small, appropriate technologies that cater for and take into consideration people's human needs. In contrast, 'big' solutions that were tied to big markets and global economic systems would only dehumanise societies.

As this book's final pages are being written, in the summer of 2020, the world is facing an unprecedented crisis. COVID-19, a novel virus originating in Wuhan, China, is ripping through populations from East to West and from North to South. As the global economy is being hit and the world is bracing itself for a recession, questions around the catastrophic impact of globalisation and theories of de-globalisation are also emerging. Schumacher's 'small is beautiful' idea couldn't sound more relevant. With the global food supply chain being susceptible to disruptive trade-related measures and restrictions, questions around food security and the resilience of the global food system are also inevitably being asked. Is this pandemic a warning that sustainable food production and consumption is possible only on a local scale? Is this the time for consumers to return to self-sufficiency and confine their purchases to produce that is sourced locally? Considering that millions of people around the world depend on international trade for their livelihoods, it's hard to imagine a new world order where international food trade takes a back seat. Yet, more than ever, this pandemic brings to the forefront the complex but essential ways in which the local and global can and should co-exist. I feel that E. F. Schumacher's *Small is Beautiful* is a timely reminder about the need for a people-centred economic system that will remain within people's control, encouraging human interaction. Does this mean that we somehow need to find ways to 'reverse' globalisation? Would that even be possible or desirable? Perhaps what the world needs is to find creative ways for the local and the global to exist in a more balanced, less harmful manner.

One of the core messages in this book concerned the need for more attention to be paid to smaller and localised development projects and initiatives. Sustainable development is a highly contingent concept and process and understanding it requires paying closer attention to the small and localised experience of the people. But small, beautiful, and local sustainable development does not have to be seen as a limiting approach because the small and the local are also open and connected, and looking for sustainable solutions through small-scale and local projects can also have a global reach. I find that Ezio Manzini's SLOC (Small, Local, Open, Connected) scenario (2007; 2011; 2015) offers an optimistic approach to achieving a sustainable yet networked and global economy and society. Manzini has also drawn inspiration from Schumacher's ideology, but in light of the new global reality, he stresses how the 'small' that Schumacher referred to can be influential as a node in a global network, while the local can be open to global flows of people, information, and ideas (Manzini 2015: 177–178). Although Manzini's SLOC scenario was developed as a model for the design of materials for social innovation and sustainability, it encapsulates an all-encompassing framework for understanding global networks and the interactions that constitute the global food system more broadly.

Manzini (2007, 2011) draws on the network society theory to conceptualise the interaction among small, local systems that reconstruct the whole, like nodes in a network. In this network, locality-oriented initiatives remain important as strategies that will enhance the self-sufficiency of the local community. Sustainable solutions should focus on and be found in the small and local places. But for these solutions to work, these small entities and their localities should remain connected to a global network. These interconnections are captured in 'creative communities' that include institutions, enterprises, non-profit organisations, and individual citizens that are acting outside of the mainstream economic models and are inventing and enhancing new socio-cultural and economic activities to support the sustainability of local communities. He gives the example of the Ainonghui Farmers' association in China that has helped small farmers sustain traditional farming and secure higher incomes by connecting with consumers in the distant big cities (Manzini 2011). Although the specific example refers to relations and interactions within the boundaries of the national, it exemplifies a form of social interchange that relies on activities based on local resources and skills that are linked into wider global networks.

Understanding how small and local systems interact and how they can contribute to sustainable solutions requires first an understanding of development on a small and local scale based on the unique experiences of small communities. Understanding the social relations and communication practices that connect communities can pave the way for identifying mechanisms and methods to connect and spread sustainable behaviours and solutions. The experiences presented in this book offer a glimpse into the possibilities that exist. Through their practices and their resistance to society's unfavourable norms and rules, the Wayanadan farmers become a source of inspiration that could influence a network's activities in a way that helps them create a more sustainable future for themselves and other networks. Through their mentality and actions, they exhibit certain characteristics associated with the concept of Manzini's 'creative communities'. These are communities that create locality-oriented initiatives, such as strategies to enhance the self-sufficiency of the local community, and try to enhance solutions to everyday life problems without waiting for wider changes in the system, such as the economy. The Wayanadan farmers are an example of a community that is challenging mainstream ways of thinking in their society in order to bring their individual interests and well-being in line with the interest of the environment. Faced with everyday life problems and questions such as the need for natural food, healthy living conditions, and sustainability, they take action in order to provide a solution. Manzini (2007) highlights the importance of support mechanisms, including information exchange and intervention by institutions, civil society organisations, or companies required for the sustainability and long-term existence of such communities, which seems to agree with the needs of communities similar to the farmers in Wayanad. It seems that the challenge is in building a stronger local network that will enable the sustainability of the place. But, for the time being at least, maintaining their connection to the global network is paramount for their growth.

Although the maintenance of small-scale farms may seem contrary to the process of globalisation and the reality of the global food market, it does not have to be. On one hand, society will be better off with productive small-scale farming that will help build strong rural economies in the Global South and allow people to remain with their families (Altieri 2009). This could not be truer for places likes Wayanad where the migration of young people to the cities or abroad in search of employment and social recognition remains one of the main challenges. This highlights the importance of creating financial incentives for younger generations and ultimately building a strong rural economy; and these financial incentives are also inextricably tied to export opportunities that provide the principal revenue stream. More importantly, the experience of the Wayanadan farmers shows that a mesh of connected localities can exist on more than just the economic level. The relationship and philosophy they share with many of their foreign buyers is a testament of the sociability of the economy and the social bonds that connect actors in the global economic system. Under the extraordinary pressures of a global crisis, such social bonds could safeguard societies and economies from disintegration. Hence, if as Manzini (2007) suggests, the small and the local acquires new meaning in the context of the network society and becomes a node in a global network, its potential to influence the global market could come in different forms, such as by spreading an ethical philosophy and environmental stewardship. The virtual interaction that took place between the farmers in Wayanad and the consumers in the UK is another example of how different localities can connect. The impact of such connections is no longer linked to the size/scale of localities alone but to the quantity and quality of the links between them. Replicating these connections should therefore become priority question and requires giving visibility to small, beautiful, and *weak* cases, relations and interactions that remain hidden and can help develop a more ethical and sustainable global food provision system. Equally paramount is the role of actors from the private sector and civil society for maintaining a creative balance between local and small-scale networks and the global food network. In this case, the communication and the relations that farmers can build with businesses and consumer groups are but one example of how these connections can materialise.

Small is Beautiful is an idea that is incarnated in the experiences of the farmers, the small businesses and the consumers that became the focus of this book. Through their practices and beliefs, these actors displayed a desire to be part of a global food economy that remains within their comprehension and creates space for closer relations and human interaction. In this case, the empowering nature of the *Small and Beautiful* is constituted also through the power of 'weak' communication. Together, the two ideas call for more attention to human relations for a sustainable future, and at the time of writing these words, as the COVID-19 pandemic unravels, these relations could not be more crucial. The concept of weak communication is an attempt to do just that. It drives our attention to the relational dimension of the economic transactions that connect farmers, businesses, and consumers and that brought to light the closer interrelation of the social, environmental, and economic aspects of sustainability. This interrelation showcases the multidimensional nature of sustainable development that depends on a combination of material assets with psychological and social resources. It also proves that development can be an 'irrational' process (Kleine 2013), such as in the case of the Wayanadan farmers who prioritise a simple life and lower income in return for environmental and biodiversity protection, or the small business owner who spends considerable time talking to and listening to the farmers to understand their needs. Understanding the social relations and communication practices that connect communities will also pave the way for identifying mechanisms and methods to connect and spread sustainable behaviours and solutions, fostering the vision of a multi-local society and a multi-level sustainable development.

The Power of Weak Communication

Ezio Manzini's SLOC scenario draws our attention to a type of sustainable development that is built on interconnections between small and local communities and their creative practices that make up the global system. To this scenario, I propose adding the 'dialogic' dimension. I use the term 'dialogic' rather than 'communicative' to emphasise the informal dialogic communication processes that were the focus of this book. Dialogic communication is mostly considered peripheral in sustainable development projects, but the experience of the Wayanadan farmers, the buyers, and the consumers showed that it is a powerful process of self-discovery that can lead to action for development in multiple ways.

Brazilian educator Paulo Freire has been a leading figure in exemplifying and celebrating the power of dialogic communication through his critical pedagogy method. In this, dialogue between the teacher and the student becomes a source of mutually acquired knowledge as both teacher and student gain a deeper comprehension of their world and the action they can take to transform it. The 'teacher-student' relationship articulated in Freire's approach has, in theory at least, informed the work of participatory development programmes (see Chap. 3), with development agencies and practitioners often assuming the role of the instructor. I advocate that the dialogic process and mutual learning of this philosophy can be extended to everyday life contexts and relations such as the trade relations explored in this book. In his reading of Freire's work, Singh (2008: 712) explains that the possibility for people to find their voice and critical consciousness (in line with Freire's *conscientização*) can be expanded to a host of other pedagogic contexts. I propose that it can be extended to other non-pedagogical contexts, such as the market. Dialogic communication that is embedded in economic transactions has the power to instil a sense of confidence that leads to action. Although interactions are undoubtedly shaped by power relationships, for example, between farmers and buyers, such power relationships can also occur respectfully as both sides grow and benefit in their respective areas of conduct. The concept of weak communication was an attempt to conceptualise the power of dialogic communication in this context. In this book I drew on the theories and insights from the field of communication for development and social change, but further conceptualisation and empirical investigation of the links between informal communication processes and sustainable development is required. We need to develop theoretical tools and identify spaces where such communication can be studied and encouraged so that a deeper and broader perspective can be gained on the ways in which communication can contribute to sustainable development through a type of collective action and movement that is small, local but also connected. Informal and dialogic communication processes in food networks bring together actors in a less formalised way than centralised and organised movements. Moreover, its effectiveness is possible to assess mainly on a small and local scale through local action. Yet, when the right connecting nodes are available, small and local action can have a bigger and more global impact. Social scientists can play a pivotal role not just in giving visibility to such communication processes but also to the nodes that can bring the various small and local actions into a bigger and more global network. Such nodes can be found in the role of the media, educational institutions, businesses, or charities and non-governmental organisations, to name but a few.

A note on the role of communication technologies in supporting informal and dialogic communication processes is essential at this point. On the one hand, the findings from the work presented in this book reinforce existing theories about development not being inherent in technology. The close relations and interpersonal communication that connect farmers with the buyers were made possible through a combination of communication tools, including visits, face-to-face meetings, mobile phones, and the internet. Yet, it was the physical visits and face-to-face encounters that have had the most lasting effect from a development point of view, with technology serving as a facilitator. In this case, the affordance of technology is in fact an affordance of the philosophy and intent of the buyers and the farmers. Along similar lines, the empowering nature of the storytelling experiment was an affordance of the participatory video production process, which encouraged dialogue among the farmers, reflexivity, and action. On the other hand, and as the farmer-consumer internet-based encounter demonstrated, the role of technology was critical in providing a platform for this interaction.

Understanding the role of communication technologies in sustainable development requires an understanding of the complex communicative channels and information flows in peoples' lives. Here, the concept of communicative ecologies provides a valuable framework (Slater 2013; Tacchi et al. 2003). Communicative ecologies have been described as 'processes that involve a mix of media, organized in specific ways, through which people connect with their social networks' (Tacchi et al. 2003: 17). The concept of communicative ecologies draws attention to the insufficiency of access to technology for development if these technologies do not fit into people's existing ecologies. Instead, technology should be placed in the context of locally significant communication processes, which includes interpersonal communication. The farmers' experience with the video experiment is a telling example of the inextricable connection between the use of a specific technology—a mobile phone camera—and the dialogic communication among the farmers. It also shows a different connection between the mobile phone and agricultural development that is materialised through the video making process and the use of the mobile phone as a creative marketing and training tool, as opposed to more common uses of mobile phones, such as the provision of information about market performance and prices. Apart from enabling a more nuanced understanding of ICTs in development, the concept of communicative ecologies offers an analytical framework for a more systematic study of information and communication technology, structures, processes, and practices in sustainable development. Having been designed to take into account relationships and contexts, it lends itself well to the study of ICTs and development in new empirical contexts, such as global food networks, and the development of new theoretical frameworks about the transmission of knowledge in a business context that is not limited to market-related information.

Business, Communication, and Sustainable Development: Towards a New Paradigm

One of the key messages reiterated through this book is the multidimensional nature of sustainable development that entails a multiplicity of strategies and new types of partnerships between the public and private sector (Van Hemelrijck 2013: 29). In the context of the Sustainable Development Goals agenda, the UN is already working together with businesses and multi-national corporations. According to a statement made by UN Secretary-General António Guterres, the private sector is a 'driving force to push governments to assume their responsibilities' (UN 2018 in Miklian 2019: 570). Yet, as Chap. 4 demonstrated, discussions on the business-development nexus are fraught with tensions and inconclusive answers. In the field of communication and development, critiques have focused primarily on political economy questions around the role of private corporations as funding mechanisms for development programmes, and the exercise of power of a small number of agencies such as the Ford and Rockefeller Foundations, the Open Society Foundation, Johns Hopkins, the Bill and Melinda Gates Foundation, and UNICEF (Thomas and van de Fliert 2014: 62). The role of the private sector is mostly seen as liable for reducing development and social change to entrepreneurialism and corporate social responsibility (Enghel 2015; Wilkins and Enghel 2013; Quarry and Ramírez 2009; Thomas and van de Fliert 2014). Discussions are also placed in the context of globalisation, capitalism, and the neoliberal agendas and capitalist values of the funders and private corporations that drive the design of development initiatives at the expense of the beneficiaries (Wilkins and Enghel 2013; in Enghel 2015: 16). Inherent in these critiques is a broader critique of market practices, which according to Escobar (2001) were imposed on development programmes through elite academies, international institutions, and western governments.

Without denying the legitimacy of these critiques, I argue that a blanket approach to business and the private sector risks missing evidence of and opportunities for communication processes in the context of business and trade relations that can contribute to development. For instance, Singh (2008) highlights the valuable role of the markets—situated in every culture and imaginary—in providing affordable communication technologies for development. More crucially, he calls for a view of the market and market exchanges from below and based on rich ethnographic data, evidence, and analysis, such as the anthropological analysis conducted by Guyer (2004) on the monetary transactions, symbols, and meanings of market exchange in Africa through history (Singh 2008: 719). The analysis of the market exchanges that bring together the Wayanadan farmers with their foreign buyers through the prism of communication for development was my own attempt to provide another viewpoint of the market from below, and to also contribute towards the idea of a new paradigm, where private enterprises gain a more permanent place in development and social change initiatives. The idea of a new paradigm was introduced by Emile McAnany (2012, 2014), based on his analysis of social entrepreneurship as a new business approach to development, where social change originates by local people with innovative ideas and is funded by private enterprises. As Noske-Turner (2015) argues, the term 'entrepreneur' is admittedly disconcerting in development circles, as it carries with it a connection with neoliberalism, which in turn contrasts inclusivity and participation. Yet, there is space in this new paradigm for slow, long-term business relations, which might lead to more sustainable social change and development. Development organisations and NGOs could embrace the role of the private sector. In our role as social scientists, we can begin by discovering and bringing more visibility to such opportunities. If, as Sedgwick (2003) suggests, knowledge is performative, this is the way to actively participate in constituting sustainable development and social change.

Depoliticising Sustainable Development: A Message from Wayanad

> Not even an hour goes by without agriculture in our life, or without spending time with the plants we love. This is regardless of the season or weather conditions. So our daily life starts and ends with our farm. Agriculture is part of our spirituality. This farm gives us pleasure, love and goodness. And we are giving the same back to the world. (Extract from a farmer's written story)

One of the main reasons for writing this book was to contribute to a more holistic understanding of sustainable development or a 'depoliticisation' of sustainable development by shifting the attention away from the 'growth ideology and the obsolete idea that growth must be placed at the core of economic and sustainability policy' (Gómez-Baggethun 2018: 73). The above quote—part of a story written by a farmer for the storytelling project—is a succinct reflection of the need for such a shift of attention.

The sense of spirituality described in this story summarises the human and social well-being dimension of sustainable development that continues to remain elusive in sustainable development theory and practice. Spirituality was a very palpable quality in Wayanad, articulated through the farmers' connection to nature; and it co-existed with an entrepreneurial spirit and profound agricultural knowledge and capacity to develop innovative solutions. At the same time, I couldn't help but notice insecurity and doubt in the farmers' own skills, as they were contemplating using the 'modern' methods used in places like Brazil, for growing pepper on concrete pillars, despite the effectiveness of their methods and the value addition these offered. Throughout the course of this research journey, it became evident that this insecurity was attributed, at least to a certain extent, to the social marginalisation of organic farmers and to the lack of opportunities for them to express their voice and be listened to by their own community and the world more broadly, with the foreign buyers offering a valuable outlet. The empowering nature of voice and listening was discussed in the previous chapters. Here, I want to emphasise the need for further research into the complex relationship between the less visible aspects of sustainable development, such as psychological resources, and how they connect to material and natural resources. Although this connection has been conceptualised through different theoretical frameworks (see Chap. 1), there seems to be a vacuum in empirical research that demonstrates the various ways in which the tangible and the intangible aspects of sustainable development blend. A focus on voice and self-expression as a form of intangible/immaterial development can provide the anchor for more systematic analysis and theorisation. Finally, I want to make a note on the significance of the farmers' indigenous knowledge for sustainable development. This knowledge concerns not just agriculture methods but a deeper and broader understanding of the spiritual and relational aspects of sustainability and development that the world can learn from. In effect, understanding, depoliticising, and achieving sustainable development will not be possible without bringing the voices and knowledge of these communities to the forefront, and this requires that powerful organisations, influential institutions, and policymakers learn how to listen to this knowledge.

In his inspiring book *Communicating Development with Communities*, Linje Manyozo makes a strong case about the need to listen as a deliberate process, seeking out those with inferior status, and understanding what these voices are all about. He sets out three forms of listening: listening to evidence, listening to ourselves, and listening as a form of speaking. All three forms require that policymakers, development organisations, and other influential and decision-making bodies step out of their expert role, engage with the empirical reality of communities, engage into a dialogic relationship with them, and make 'listening' their form of 'speaking' (Manyozo 2017: 129–135). I hope that this book is a step towards this direction.

References

Altieri, M. A. (2009). The ecological impacts of large-scale agrofuel monoculture production systems in the Americas. *Bulletin of Science Technology Society, 29*(3), 236–244. https://doi.org/10.1177/0270467609333728.

Carpiano, R. M. (2009). Come take a walk with me: The "go-along" interview as a novel method for studying the implications of place for health and well-being. *Health & Place, 15*(1), 263–272. https://doi.org/10.1016/j.healthplace.2008.05.003.

Ciegis, R., Ramanauskiene, J., & Martinkus, B. (2009). The concept of sustainable development and its use for sustainability scenarios. *Engineering Economics, 62*(2), 28–37.

Dimitrov, D. (2010). *The paradox of sustainability definitions, APIRA conference.* Available online: http://apira2010.econ.usyd.edu.au/conference_proceedings/APIRA-2010-073-Dimitrov-The-paradox-of-sustainability-definitions.pdf. Accessed 20 Apr 2020.

Enghel, F. (2015). Towards a political economy of communication in development? *Nordicom Review: Nordic Research on Media & Communication, 36*, 11–24.

Escobar, A. (2001). *Encountering development: The making and unmaking of the third world.* Princeton: Princeton University Press.

Gibson, R. B. (2006). Beyond the pillars: Sustainability assessment as a framework for effective integration of social, economic and ecological considerations in significant decision-making. *Journal of Environmental Assessment Policy and Management, 8*(3), 259–280.

Gómez-Baggethun, E. (2018). Sustainable development. In A. Kothari, A. Salleh, A. Escobar, F. Demaria, & A. Acosta (Eds.), *Pluriverse: A post-development dictionary* (pp. 71–73). New Delhi: Tulika Books.

Guyer, J. (2004). *Marginal gains: Monetary transactions in Atlantic Africa.* Chicago, IL: University of Chicago Press.

Kleine, D. (2013). *Technologies of choice? ICTs, development, and the capabilities approach.* Cambridge, MA: MIT Press.

Manyozo, L. (2017). *Communicating development with communities*. London: Routledge.

Manzini, E. (2007). Design research for sustainable social innovation. In R. Michel (Ed.), *Design research now. Board of International Research in design* (pp. 233–245). Basel: Birkhäuser.

Manzini, E. (2011). *The new way of the future: Small, local, open and connected, social space* (pp. 100–105). Available at https://ink.library.smu.edu.sg/lien_research/75.

Manzini, E. (2015). *Design, when everybody designs: An introduction to Design for Social Innovation*. Cambridge, MA: MIT press.

McAnany, E. G. (2012). *Saving the world: A brief history of communication for development and social change*. Urbana: University of Illinois Press.

McAnany, E. G. (2014). Economics and communication for development and social change. In K. G. Wilkins, T. Tufte, & R. Obregon (Eds.), *Handbook of development communication and social change* (pp. 242–258). London: Wiley-Blackwell.

Miklian, J. (2019). The role of business in sustainable development and peace-building: Observing interaction effects. *Business & Politics, 21*(4), 569–601. https://doi.org/10.1017/bap.2019.28.

Miller, T. R., Wiek, A., Sarewitz, D., Robinson, J., Olsson, L., Kriebel, D., & Loorbach, D. (2014). The future of sustainability science: A solutions-oriented research agenda. *Sustainability Science, 9*(2), 239–246. https://doi.org/10.1007/s11625-013-0224-6.

Noske-Turner, J. (2015) Left for dust: Innovation and C4D. In *Mobilising media for sustainable outcomes in the Pacific region*. Available at: https://sites.google.com/site/mobilisingmediainthepacific/research-from-the-field/leftfordustinnovationandc4d. Accessed 7 Apr 2020.

Quarry, W., & Ramírez, R. (2009). *Communication for another development: Listening before telling*. New York: Zed Books.

Schumacher, E. F. (1973). *Small is beautiful: A study of economics as if people mattered*. London: Blond and Briggs.

Sedgwick, E. K. (2003). *Touching feeling: Affect, pedagogy, performativity*. Durham/London: Duke University Press.

Servaes, J., & Lie, R. (2013). Sustainable social change and communication. *Communication Research Trends, 32*(4), 1–43.

Singh, J. P. (2008). Paulo Freire: Possibilities for dialogic communication in a market-driven information age. *Information, Communication & Society, 11*(5), 699–726. https://doi.org/10.1080/13691180802124518.

Slater, D. (2013). *New media, development and globalization: Making connections in the global south*. Cambridge, UK: Polity Press.

Tacchi, J., Slater, D., & Hearn, G. (2003). *Ethnographic action research handbook*. New Delhi: UNESCO.

Thomas, P. N., & van de Fliert, E. (2014). *Interrogating the theory and practice of communication for social change: The basis for a renewal.* London: Palgrave Macmillan.

Trencher, G. P., Yarime, M., & Kharrazi, A. (2013). Co-creating sustainability: Cross-sector university collaborations for driving sustainable urban transformations. *Journal of Cleaner Production, 50*(1), 40–55. https://doi.org/10.1016/j.jclepro.2012.11.047.

Trott, C. D., Weinberg, A. E., & Sample McMeeking, L. B. (2018). Prefiguring sustainability through participatory action research experiences for undergraduates: Reflections and recommendations for student development. *Sustainability, 10*, 1–21. https://doi.org/10.3390/su1.

United Nations Office of the High Commissioner on Human Rights (UNHRC). (2018, September 18). *Report of the independent international fact-finding mission on Myanmar.* A/HRC739/64. New York: United Nations.

Van Hemelrijck, A. (2013). Powerful beyond measure? Measuring complex systemic change in collaborative settings. In J. Servaes (Ed.), *Sustainability, participation and culture in communication: Theory and praxis* (pp. 25–58). Bristol: Intellect.

Vandecandelaere, E., Arfini, F., Belletti, G., & Marescotti, A. (2009). *Linking people, places and products.* A guide for promoting quality linked to geographical origin and sustainable geographical indications; FAO: Rome, Italy. Available online: http://www.fao.org/docrep/013/i1760e/i1760e00.pdf. Accessed 19 Apr 2020.

Wilkins, K. G., & Enghel, F. (2013). The privatization of development through global communication industries: Living proof? *Media, Culture and Society, 35*(2), 165–181. https://doi.org/10.1177/0163443712468606.

Index[1]

[1] Note: Page numbers followed by 'n' refer to notes.

M. Touri, *Sustainable Development and Communication in Global Food Networks*, https://doi.org/10.1007/978-3-030-46119-5

The manufacturer's authorised representative in the EU is Springer Nature Customer Service Centre GmbH, Europaplatz 3, 69115 Heidelberg, Germany. If you have any concerns regarding our products, please contact ProductSafety@springernature.com

Printed and bound by CPI Group (UK) Ltd, Croydon, CR0 4YY
30/09/2026
02225644-0001